The Agile Landscape

A Context-Driven Guide to Modern Agile Project Management

M. Rafati

The Agile Landscape: A Context-Driven Guide to Modern Agile Project Management

Published by Hybrid Exploratory Approach
Ottawa, Ontario, Canada
www.h-ex-a.com

Hybrid Exploratory Approach is a trade name of Mohammad Rafati Shaldehi.

This book is published under the author name: M. Rafati

ISBN: 978-1-0675713-0-6 (Paperback)
ISBN: 978-1-0675713-1-3 (Hardcover)
ISBN: 978-1-0675713-2-0 (E-book)

Trademarks

Disclaimer

The information provided in this book is for educational and informational purposes only. The author and publisher assume no responsibility for errors, inaccuracies, or omissions. The strategies outlined in this book may not be suitable for every situation. Professional advice should be sought where appropriate.

Library and Archives Canada Cataloguing in Publication
Rafati Shaldehi, Mohammad
The Agile Landscape: A Context-Driven Guide to Modern Agile Project Management / Rafati Shaldehi, Mohammad
Ottawa: Hybrid Exploratory Approach, 2026.
Includes bibliographical references and index.
Identifiers: ISBN: 978-1-0675713-0-6 (Paperback) | ISBN: 978-1-0675713-1-3 (Hardcover) | ISBN: 978-1-0675713-2-0 (E-book)
Subjects: LCSH: Project management. | Agile project management. | Agile software development.
Classification: LCC T56.8 .R34 2026 | DDC 658.4/04—dc23

First Edition: May 2026 Printed in Canada

Preface

For most of my career, I have lived and breathed the world of predictive project management. Working in the Architecture and Construction fields, I was trained in a landscape where "measure twice, cut once" isn't just a saying—it is a financial and safety requirement. In this environment, long-range planning, fixed sequences, and robust risk management are the bedrock of success. For years, the traditional Waterfall approach wasn't just a choice; it was the only logical way to operate.

However, during recent years, I began to notice a seismic shift. The boundaries between "predictive" and "adaptive" didn't just blur; they began to merge. Even in the historically rigid sectors of construction and design, hybrid approaches emerged. I saw firms struggling not just with schedules, but with *complexity*—managing rapid digital innovation alongside physical constraints, all while navigating a world that had moved from volatile to **BANI** (Brittle, Anxious, Nonlinear, and Incomprehensible).

As I navigated this shift, I realized that many project managers—particularly those rooted in predictive landscapes—were deeply puzzled. To someone used to the clarity of a Gantt chart, the "Agile Landscape" can feel like a chaotic forest of confusing terms, overlapping frameworks, and contradictory advice.

This confusion has only deepened in 2026. The landscape is no longer just about Scrum or Kanban; it now includes **Agentic AI** teammates, **Product-Led** methodologies like *Shape Up*, and structural patterns like *Team Topologies*. Furthermore, the release of the **PMBOK Guide – Eighth Edition** has fundamentally changed the conversation, validating that "process" and "agility" are not enemies, but partners in value delivery.

There was a clear and urgent need for a guide that didn't just advocate for one "best" framework, but instead mapped the entire territory—from the traditional to the futuristic.

This realization led to the creation of *The Agile Landscape: A Context-Driven Guide to Modern Agile Project Management.*

This book is written for two distinct groups of professionals:

1. **The "Predictive" Project Manager:** Those who, like me, come from traditional backgrounds and are looking for a clear, jargon-free way to understand how to deploy Agile and Hybrid approaches within a structured governance model.
2. **The "Agile" Practitioner:** The Scrum Masters and Coaches who may be experts in one specific framework but lack a vast perspective of the emerging tools—like **Flight Levels** and **AI-driven governance**—that are defining the next decade of work.

By understanding the historical origins, the technical mechanics, and the strategic trade-offs of each methodology, you will be empowered to move beyond "copy-pasting" frameworks. Instead, you will learn to **Tailor** your approach, selecting the right tools for your specific organizational context.

My hope is that this book serves as your compass, helping you navigate the complexity of modern practice and find the most efficient path toward delivering value, regardless of the field—or the future—you work in.

M. Rafati

Table of Contents

List of Tables and Figures

Introduction: Navigating the Landscape

Agile is no longer a "new" way of working. In the decades since the signing of the Agile Manifesto in 2001, it has evolved from a rebellious software movement into a dominant influence on modern organizational management (Dingsøyr et al., 2012). Yet, despite its widespread adoption, many organizations find themselves *doing* Agile without *becoming* truly agile. They adopt the visible rituals—the stand-ups, the sticky notes, and the two-week sprints—but fail to achieve the responsiveness, technical excellence, and cultural health that the movement originally promised.

The primary reason for this disconnect is a lack of contextual awareness. We have moved from a world of too few options to a world of too many. Today's practitioners face a crowded and often confusing landscape of frameworks and methods: Scrum, Kanban, XP, SAFe, LeSS, Disciplined Agile, and the emerging world of AI-driven delivery. Without a clear way to navigate this terrain, teams and leaders frequently adopt a framework because it is popular, mandated, or familiar—only to discover later that it is fundamentally mismatched to their team size, technical constraints, regulatory environment, or organizational culture.

The Purpose of This Book

The Agile Landscape: A Context-Driven Guide to Modern Agile Project Management was written to be that guide. This book is built on a single core principle: **Context Counts**. There is no "best" Agile framework. There is only the right approach for a specific situation, at a specific point in time.

The goal of this book is not to promote or defend any single methodology. Instead, it aims to provide a clear map of the modern Agile landscape as it stands in 2026. We will explore not only how different frameworks work, but why they exist, the problems they were designed to solve, and the conditions under which they are likely to succeed—or fail. Throughout the book, the emphasis remains on informed choice and professional judgment rather than blind adherence to prescribed practices.

How This Book Is Structured

To support this navigation mindset, the book is organized into six parts, each addressing a distinct aspect of Agile practice:

- **Part I: The Agile Foundation (The "Why")** establishes the psychological and philosophical bedrock of Agile thinking. It examines the shift from a VUCA world (Volatility, Uncertainty, Complexity, Ambiguity) to a **BANI** world (Brittle, Anxious, Nonlinear, Incomprehensible), and the mindset required to operate effectively within it.
- **Part II: Lean and the Primary Agile Frameworks (The "How")** explores the foundational approaches that shape most modern Agile practice. This includes Lean thinking, Scrum, Kanban, and the rigorous technical discipline of Extreme Programming (XP).

- **Part III: The Agile Family and Hybrid Approaches** examines specialized and historically significant methodologies such as FDD, DSDM, and Crystal. These frameworks reveal important lessons about adaptation, governance, and context that continue to influence Agile practice today.
- **Part IV: Scaling Agile for the Enterprise** addresses the challenge of coordination across multiple teams and complex organizations. It contrasts heavyweight scaling frameworks such as SAFe with descaling approaches like LeSS, highlighting the organizational assumptions behind each.
- **Part V: The Practitioner's Toolkit** focuses on the practical skills required for day-to-day effectiveness, including backlog refinement, estimation, prioritization, and value-based decision-making.
- **Part VI: Governance, Leadership, and the Future** looks beyond team practices to the organizational shell surrounding Agile work. It explores the evolution of the PMO into the **Value Management Office (VMO)**, the shift to predictive governance, and the revolutionary impact of **Agentic AI** on how work is organized and managed.

Using the Framework Operational Profiles

A distinguishing feature of this book is the **Framework Operational Profile** included at the end of each primary delivery and scaling framework chapter (Chapters 4–6, 8–13, and 15–20). These profiles provide standardized, side-by-side snapshots of execution models, which **Appendix A** consolidates into a single comparative matrix for easy side-by-side analysis, while **Appendix B** offers a decision guide designed to help readers identify a sensible starting point based on their current context rather than framework popularity.

A Note to the Reader

Whether you are a predictive project manager adapting to the 2026 release of the PMBOK Guide – Eighth Edition, a Scrum Master seeking stronger technical practices, or an executive trying to understand how to govern and fund Agile work in the age of AI, this book is written for you.

The Agile landscape is not static. Frameworks evolve, organizations change, and constraints shift over time. This book does not offer a fixed set of rules to follow. Instead, it offers a compass—one that encourages reflection, adaptation, and the continuous, disciplined pursuit of a better way of working. True agility is not achieved by following a framework faithfully, but by developing the judgment to choose, adapt, and evolve your own Way of Working.

Part I: The Agile Foundation (The "Why")

Chapter 1: The Context of Agile

1.1. Why Agile? Thriving in a VUCA World

Agile did not emerge as a mere software trend—it emerged as a necessary response to a fundamental shift in how value is delivered in the modern economy. In today's "Knowledge Age," success is defined not by predictability, but by the ability to continuously sense and respond to change.

This shift challenges traditional predictive management models, which assume stability, repeatability, and upfront certainty.

This transition is driven by a landscape defined as **VUCA** According to the **Seventh Edition of the *PMBOK Guide* and the *Agile Practice Guide* (Project Management Institute [PMI], 2021)**, project environments are now characterized by:

- **Volatility:** The extreme speed and nature of change. A market lead can vanish overnight due to a single technological breakthrough.
- **Uncertainty:** The lack of predictability. Even with massive data, the link between cause and effect is often hidden until after an event occurs.
- **Complexity:** The staggering number of moving parts and interdependencies. In complex systems, solving one problem often creates three new ones.
- **Ambiguity:** The "haziness" of reality. It is the lack of clarity about how to interpret a situation; the "who, what, and why" are all subject to conflicting interpretations.

Who Must Navigate This Landscape?

This book is designed for those standing at the intersection of these forces:

- **Project Managers** transitioning from "controlling" to "facilitating."
- **Product Leaders** needing to validate ideas before over-investing.
- **Executives** seeking organizational resilience.
- **Teams** struggling with shifting priorities and "invisible" work.

In Vuca world, the cost of a "wrong" plan is high, but the cost of a "slow" plan is higher. Agile minimizes this risk by shortening the feedback loop.

1.1.1. The Shift to BANI

While VUCA describes a world that is difficult to predict, some experts claim that the modern landscape often feels impossible to comprehend. Futurist Jamais Cascio (2020) argues that we have moved from a VUCA world to a **BANI** world:

- **Brittle:** Systems that appear strong but shatter suddenly (e.g., global supply chains). The Agile response is *Resilience*—building slack and redundancy into the system.
- **Anxious:** The sense of helplessness created by data overload. The Agile response is *Empathy* and psychological safety.
- **Nonlinear:** Cause and effect are disconnected (e.g., a small code change crashing a global platform). The Agile response is *Adaptability* and rapid feedback loops.
- **Incomprehensible:** Events that defy logic and linear cause-and-effect reasoning. In the 2026 context, however, we are seeing a shift in how we respond. While the world may be incomprehensible to human cognition alone, it is becoming navigable through **Agentic Stability**. By leveraging AI agents that can process vast signals and simulate infinite scenarios, we augment our intuition with data-driven foresight. The Agile response is no longer just "transparency"—it is **Augmented Intelligence**, using AI to find patterns in the chaos that the human mind cannot see.

BANI validates why rigid plans fail: they are often *brittle*—optimized for efficiency but lacking the resilience to withstand chaos; which is often Recurring in the predictive projects.

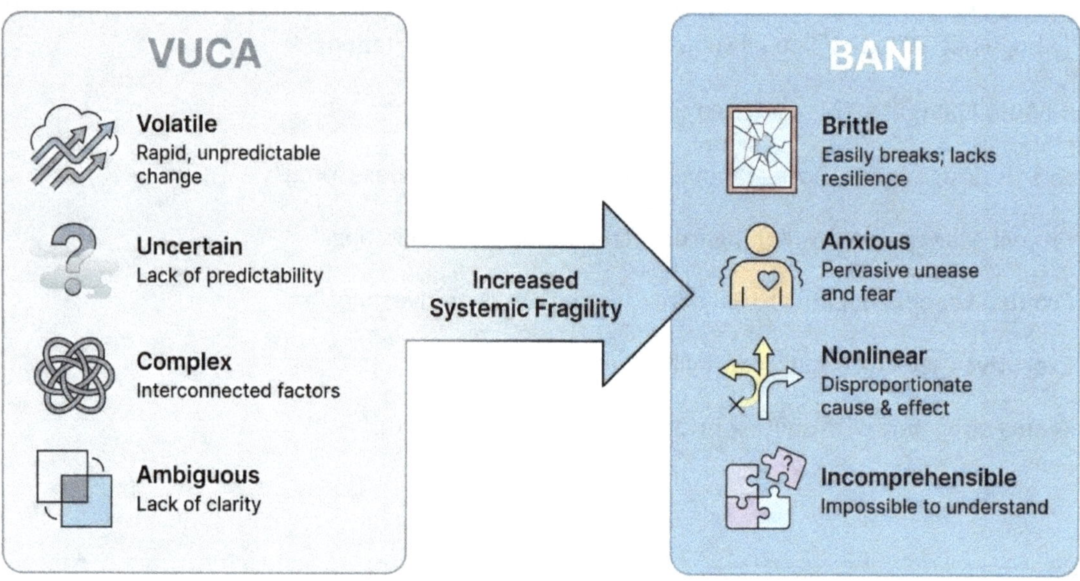

Figure 1-1: The Evolutionary Leap: VUCA to BANI

1.2. The Stacey Matrix & Cynefin Framework: Knowing When Agile Works

Agile is not a "silver bullet." **The Seventh Edition of the *PMBOK Guide*** (PMI, 2021, Section 2.3) emphasizes that the development approach should be tailored to the project's level of uncertainty. To choose the right path, we use two primary maps: **The Stacey Matrix** (Stacey, 1996), which helps us categorize project characteristics, and **The Cynefin Framework**, which dictates our decision-making mode.

The Stacey Matrix: Assessing the Project

The Stacey Matrix plots projects on two axes: **Requirements Uncertainty** (What are we building?) and **Technical Uncertainty** (How are we building it?).

- **Simple:** Requirements are clear and technology is known. (Use Predictive/Waterfall).
- **Complicated:** Requires expert analysis to find the answer. (Use Predictive or Hybrid).
- **Complex:** The "Agile Sweet Spot." Requirements and technology are in flux. The only way to succeed is to experiment.
- **Chaotic:** Represents a project state where there is no agreement and no technical certainty; usually a state to be avoided or deliberately stabilized and exited.

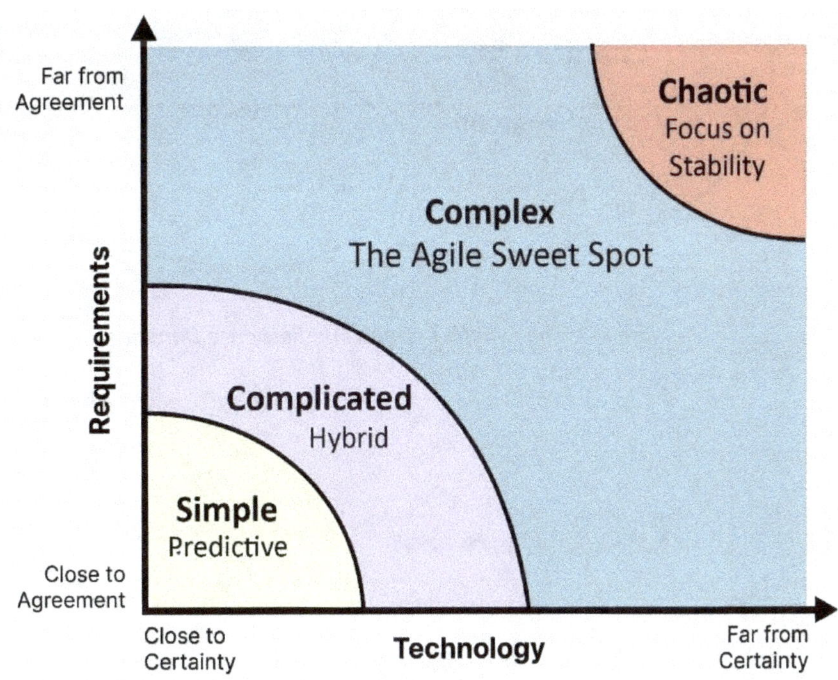

Figure 1-2: The Stacey Matrix – Assessing Project Complexity

The Cynefin Framework: Defining the Response

Created by Dave Snowden, the Cynefin framework (Snowden & Boone, 2007) helps leaders identify how to act based on the domain they are in:

- **Clear (Simple):** *Sense - Categorize - Respond.* Use Best Practices.
- **Complicated:** *Sense - Analyze - Respond.* Use Good Practices.
- **Complex:** *Probe - Sense - Respond.* Use **Emergent Practices** (The primary domain where Agile approaches are most effective).
- **Chaotic:** Represents a decision context requiring immediate action to stem the bleeding. *Act - Sense - Respond.* Use Novel Practices.

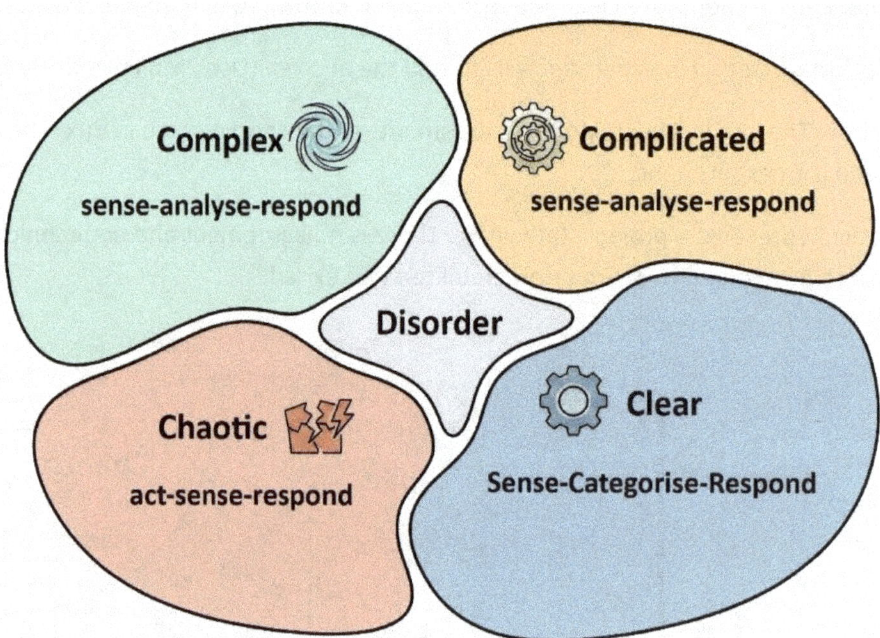

Figure 1-3: The Cynefin Framework: Response Domains

1.3. The Agile Manifesto: The Four Values and Twelve Principles

While the Manifesto was written in 2001, it remains the "constitutional document" for the most important approaches discussed in this book. However, these values manifest differently across the landscape. For instance: **Scrum** emphasizes the interaction of roles, **Kanban** emphasizes the flow of the solution, and **SAFe** applies these values to the enterprise hierarchy.

The 4 Values

1. **Individuals and interactions** over processes and tools.
2. **Working software (or solutions)** over comprehensive documentation.
3. **Customer collaboration** over contract negotiation.
4. **Responding to change** over following a plan.

The 12 Principles

These principles operationalize the four values and serve as the common foundation for all Agile, Lean, and scaling frameworks discussed in this book. While originally written for software development, these principles have since been successfully applied across product development, services, and knowledge work:

1. **Customer Satisfaction:** Highest priority is to satisfy the customer through early and continuous delivery of valuable software or solutions.
2. **Welcome Change:** Welcome changing requirements, even late in development. Agile processes harness change for the customer's competitive advantage.
3. **Frequent Delivery:** Deliver working software frequently, from a couple of weeks to a couple of months, with a preference to the shorter timescale.
4. **Collaboration:** Business people and developers must work together daily throughout the project.
5. **Support and Trust:** Build projects around motivated individuals. Give them the environment and support they need, and trust them to get the job done.
6. **Face-to-Face Conversation:** The most efficient and effective method of conveying information to and within a development team is face-to-face conversation.
7. **Working Solutions:** Working software (or solutions) is the primary measure of progress.
8. **Sustainable Development:** Agile processes promote sustainable development. The sponsors, developers, and users should be able to maintain a constant pace indefinitely.
9. **Technical Excellence:** Continuous attention to technical excellence and good design enhances agility.
10. **Simplicity:** Simplicity—the art of maximizing the amount of work not done—is essential.
11. **Self-Organizing Teams:** The best architectures, requirements, and designs emerge from self-organizing teams.
12. **Reflect and Adjust:** At regular intervals, the team reflects on how to become more effective, then tunes and adjusts its behavior accordingly.

The **Agile Practice Guide** (PMI, 2021) groups these into four main themes:

- **Satisfying the Customer:** Through early and continuous delivery of valuable work.
- **Welcoming Change:** Harnessing change for the customer's competitive advantage.
- **Frequent Delivery:** Shortening the "timebox" to increase feedback.
- **Self-Organizing Teams:** Providing the environment and trust that motivated individuals require to succeed.

Understanding this context explains *why* Agile works; adopting it, however, requires a fundamental shift in mindset, leadership, and culture—concepts explored in the next chapter.

To support practical understanding and comparison, each framework chapter in this book concludes with an **Operational Profile**. This standardized summary includes a comparative table designed to capture how the methodology works in practice and a **Framework Snapshot infographic**. These profiles are not prescriptions or maturity checklists; they are intended to help readers compare frameworks objectively and choose approaches based on context rather than dogma.

> **Note:** While most chapters feature both elements, please note that the summary table is excluded for **Lean**, **The unFix Model**, and **Agentic AI** to better reflect their fluid or emerging nature.

Chapter 2: The Core Agile Mindset

2.1. Being Agile vs. Doing Agile: The Cultural Shift

A common failure point in organizational transformations is the tendency to adopt the "mechanics" of Agile without adopting the "mindset." This is the distinction between **Doing Agile** and **Being Agile**.

- **Doing Agile:** Focuses on the surface-level implementation of ceremonies, roles, and artifacts (e.g., holding a Daily Stand-up or using a Kanban board). In this state, teams often still follow a predictive, "command-and-control" philosophy under the guise of Agile labels.
- **Being Agile:** Represents a fundamental shift in values and belief systems. It is characterized by an internal commitment to transparency, a relentless focus on customer value, and a willingness to fail fast in order to learn.

The Project Management Institute's **Agile Practice Guide** (2021) emphasizes that Agile is a mindset defined by values, guided by principles, and manifested through many different practices. If the mindset is absent, the practices become "**empty rituals**" that provide little to no business agility or measurable value.

2.2. The Evolution of Leadership: From Control to Catalyst

Leadership within the Agile landscape is not a static set of traits; it is an evolving response to complexity. To understand modern practice, we must trace the shift from the industrial mindset of the past to the integrated, value-driven mindset of the current era.

1. **The Legacy: Directing and Controlling**

In traditional project management, the leader is the "Commander." This model assumes that the leader possesses the most knowledge and that success is achieved through strict adherence to a plan.

- **Structure:** Top-down hierarchy.
- **Focus:** Task management, individual accountability, and variance reduction.
- **The Goal:** Predictability through authority.

2. **The Agile Shift: Servant Leadership (PMBOK 7)**

As organizations realized that complexity cannot be managed by a single "expert," the model inverted. **PMBOK 7 (Section 3.1)** codified this by moving from "Command and Control" to **"Coordinate and Collaborate."** This introduced **Servant Leadership** as the standard for Agile roles (Scrum Masters, Product Owners, or RTEs).

- **Removing Impediments:** The leader's primary job is to clear the path.
- **Facilitation over Direction:** Guiding the team toward a goal without dictating the "how."

- **Coaching and Mentoring:** Focusing on the growth and well-being of team members. As Lyssa Adkins (2010) emphasizes, the agile coach must transition through multiple roles—teacher, mentor, and conflict navigator—to help the team reach its own path to high performance

- **The Goal:** Empowerment and psychological safety.

3. **The Modern Frontier: Catalytic & Adaptive Leadership (PMBOK 8)**

The **PMBOK 8** era recognizes that serving the team is no longer enough in a hyper-connected, AI-augmented ecosystem. We have moved from "Serving the Team" to **"Architecting the Environment."** Leadership is now viewed as a **Catalytic Function**—an agent that triggers change and optimizes the flow of value across the entire enterprise.

- **From Shielding to Integrating:** While a Servant Leader "shields" the team from noise, a **Catalytic Leader** integrates the team into the organizational strategy, ensuring they are not just "fast," but "aligned."

- **Adaptive Governance:** Instead of just removing blockers, leaders proactively design systems (guardrails) that allow for autonomous decision-making at the edge.

- **Value-Driven Influence:** Leadership is no longer a role but a distributed competency. PMBOK 8 emphasizes that influence replaces authority entirely; the leader's job is to catalyze the team's ability to sense and respond to market shifts.

4. **Summary of the Leadership Evolution**

Table 2-1: Summary of the Leadership Evolution

Era	Leadership Style	Core Philosophy	Key PMBOK Reference
Traditional	**Command & Control**	Manage the work	Pre-PMBOK 7
Early Agile	**Servant Leadership**	Support the people	PMBOK 7
Modern Agile	**Catalytic & Adaptive**	Optimize the System	PMBOK 8

Context-Driven Note: In *The Agile Landscape*, we must recognize that a leader may need to pivot between these styles depending on the "weather." A crisis might require more direction, while a high-performing team requires a purely catalytic approach.

2.3. Empiricism, Systems Thinking, and Continuous Learning

To navigate complexity, an Agile mindset relies on three intellectual pillars that ensure the organization remains grounded in reality.

2.3.1. Empiricism (The Three Pillars)

Empiricism allows teams to respond effectively in volatile, uncertain, complex, and ambiguous environments. It asserts that knowledge comes from experience and making decisions based on what is known. It requires:

1. **Transparency:** All aspects of the process must be visible to those responsible for the outcome.
2. **Inspection:** The artifacts and progress must be inspected frequently to detect variances.
3. **Adaptation:** If a process or outcome is found to be outside acceptable limits, the process must be adjusted immediately.

2.3.2. Systems Thinking

Derived from Lean philosophy, **Systems Thinking** encourages leaders to view the organization as an interconnected whole rather than a collection of silos. It teaches that optimizing a single department (e.g., making the coding team "faster") may actually harm the overall system if it creates a bottleneck in testing or deployment.

2.3.3. Continuous Learning

Agile teams treat every iteration as a learning experiment. This involves the **PDCA (Plan-Do-Check-Act)** cycle, ensuring that the team is constantly evolving their technical skills and their process.

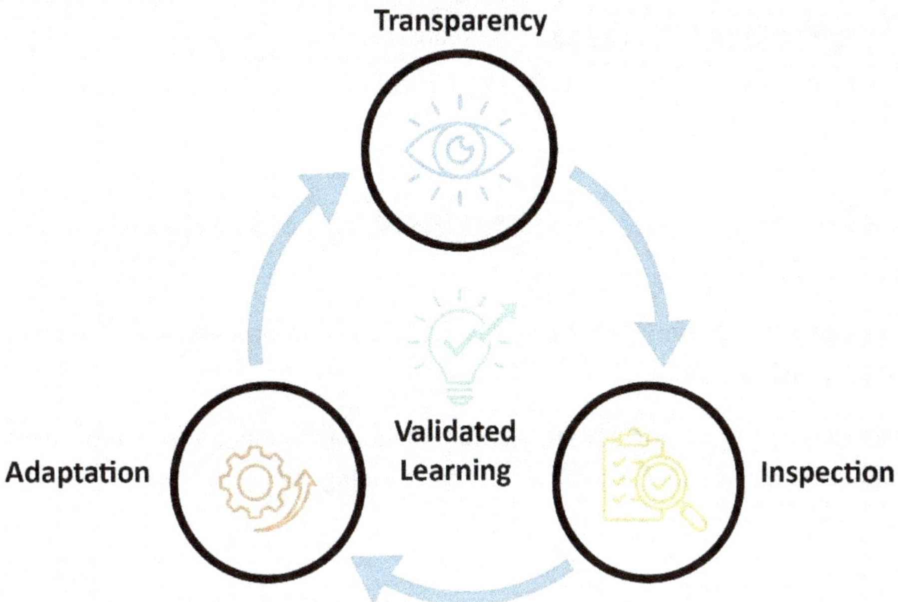

Figure 2-1: The Empirical Engine: The Three Pillars

2.4. Psychological Safety and High-Performing Teams

The most sophisticated Agile framework will fail if the environment in which it operates is governed by fear. According to Google's Project Aristotle (2015)—a landmark study on team effectiveness—and contemporary Agile research, **psychological safety** is the single most important factor in team performance.

What is Psychological Safety?

Psychological safety is the shared belief that a team is safe for interpersonal risk-taking. In a psychologically safe environment, team members feel comfortable:

- Admitting mistakes without fear of punishment.
- Asking "stupid" questions.
- Challenging the status quo or the opinions of senior leaders.
- Sharing "half-baked" ideas for innovation.

When psychological safety is absent, teams hide problems instead of exposing them. This behavior directly undermines transparency. Without transparency, inspection and adaptation—the core mechanisms of empiricism—cannot function. Psychological safety is therefore not a "soft" or optional concept; it is a foundational prerequisite for effective Agile execution.

2.4.1. Neurodiversity and Cognitive Safety

Psychological safety must extend to *cognitive* safety. A significant percentage of high-performing technical talent identifies as neurodivergent. Traditional Agile practices—like the high-pressure, verbal, face-to-face Daily Scrum—can be exclusionary for these team members.

Modern Adaptation:

- **Asynchronous Stand-ups:** Allow updates via text (Slack/Teams) to reduce social anxiety and improve focus.
- **"Camera-Optional" Culture:** Acknowledge that for some, processing visual data drains the energy needed for problem-solving.
- **Written Pre-reads:** Distribute materials before planning meetings to allow for "deep processing" time, rather than demanding on-the-spot brainstorming.

2.5. The Fixed vs. Growth Mindset

Drawing from the landmark research of Carol Dweck (2006), an Agile practitioner must intentionally cultivate a **Growth Mindset**.

Table 2-2: The Shift from a Fixed Mindset to an Agile Growth Mindset

Characteristic	Fixed Mindset (Traditional)	Growth Mindset (Agile)
Challenges	Avoids them to stay "safe."	Embraces them as a chance to learn.
Obstacles	Gets defensive or gives up.	Persists in the face of setbacks.
Effort	Seen as fruitless or a sign of weakness.	Seen as the path to mastery.
Criticism	Ignores useful negative feedback.	Learns from criticism.
Success of Others	Feels threatened.	Finds lessons and inspiration.

Part II: Lean and the Primary Agile Frameworks (The "How")

Chapter 3: Lean Thinking — The Foundation of Agile

3.1. Lean Principles and Origins

Lean Thinking originated with the **Toyota Production System (TPS)** in post-WWII Japan. While traditional Western manufacturing focused on "economies of scale" (mass production), Toyota focused on "economies of flow."

The **PMBOK Seventh Edition** integrated Lean as a method for optimizing the "Value Delivery Office" (VDO) through a principle-based lens. However, the **Eighth Edition** evolves this further, positioning Lean not just as a tool, but as the foundational architecture for the **Value Delivery Ecosystem**.

Lean is defined by five core principles popularized by Womack and Jones (2003), which modern practitioners now apply to complex, AI-augmented workflows:

1. **Identify Value:** Value is defined by the customer's needs. If the customer wouldn't pay for a specific activity, it is potentially waste.
2. **Map the Value Stream:** Identify every step in the process required to deliver the product, from concept to launch.
3. **Create Flow:** Ensure that the product moves through the value stream without interruptions, batches, or bottlenecks.
4. **Establish Pull:** Nothing is built until the downstream customer signals a need for it (Just-In-Time).
5. **Seek Perfection:** Continuously improve the system (**Kaizen**) to eliminate waste and increase quality.

These principles are not implemented directly; frameworks operationalize them in different ways.

3.2. Value Streams, Flow, and Waste Reduction

To master the Agile landscape, a practitioner must develop "Lean Eyes"—the ability to see work as a continuous flow and identify the friction points that slow it down.

3.2.1. The Value Stream

In knowledge work, the value stream is often invisible, consisting of information hand-offs between designers, developers, and stakeholders. **Value Stream Mapping (VSM)** is a critical tool used to visualize these steps. While PMBOK 7 used VSM to measure the ratio of "Value-Add Time" to "Wait Time," **PMBOK 8** elevates this to **Value Stream Orchestration (VSO)**—the active management of flow across multiple interconnected teams.

3.2.2. The Three Types of Waste (Muda, Mura, Muri)

In the Lean philosophy, waste is categorized into three specific forms that inhibit flow. In the modern landscape, these are seen as the primary "pollutants" of an Agile ecosystem:

- **Muda (Waste):** Activities that consume resources but add no value. Lean identifies **eight wastes** (often remembered by the acronym **DOWNTIME** or **TIMWOOD**): Defects, Overproduction, Waiting, Non-utilized talent, Transportation, Inventory, Motion, and Extra-processing. The specific acronym is less important than developing the ability to recognize waste and improve flow in context.
- **Mura (Unevenness):** Inconsistency in the workload. When a team is idle on Monday but overwhelmed on Friday, the "unevenness" creates stress and quality issues.
- **Muri (Overburden):** Pushing people or systems beyond their natural limits. In a modern context, this includes **Cognitive Overload**, where teams are overwhelmed by too many tools or complex processes.

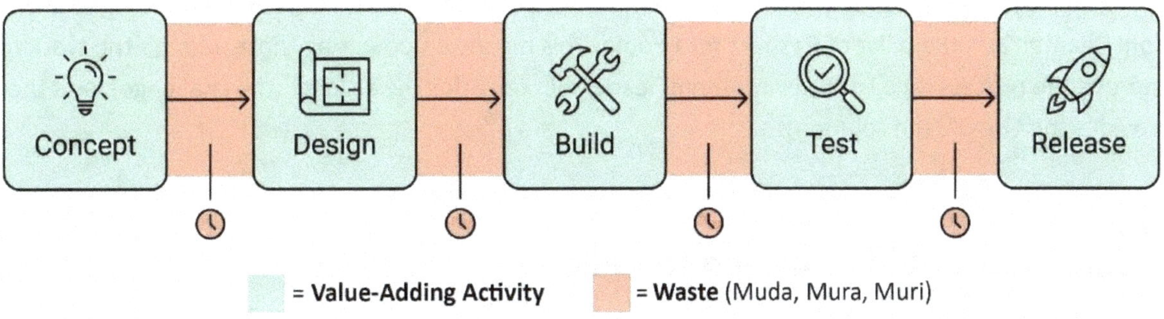

Figure 3-1: Lean Value Stream: Value vs. Waste

3.2.3. The Evolution of Waste: Carbon and Sustainability

In the modern era, Lean's definition of waste has expanded to include environmental impact. **Green Software Engineering** principles treat carbon emissions as a defect.

- **Carbon Efficiency:** Writing code that consumes less electricity.
- **Carbon Awareness:** Scheduling heavy workloads (like AI model training) when the energy grid is powered by renewables.
- **Software Carbon Intensity (SCI),** formalized in the **ISO/IEC 21031:2024 standard:** A metric used to score a product's carbon footprint per user. Just as we track "bugs per release," modern teams track "emissions per feature."

3.3. Lean's Influence on Modern Agile Frameworks

Lean Thinking is the "connective tissue" of the Agile landscape. Its influence is found in almost every framework:

- **Scrum:** Uses the Lean concept of "Timeboxing" and the "Definition of Done" to ensure quality at the source.
- **Kanban:** Is a direct implementation of Lean's "Pull" system and "Limit Work in Progress (WIP)" principles.
- **SAFe (Scaled Agile Framework):** Is built upon the **SAFe House of Lean**, which emphasizes "Flow" and "Relentless Improvement" as the pillars of enterprise agility.
- **XP (Extreme Programming):** Applies Lean's "Simplicity" and "Eliminating Waste" to the code itself through refactoring and TDD.

3.3.1. The Concept of "Respect for People"

A frequent misconception is that Lean is only about efficiency. Central to Lean (and the Agile mindset from Chapter 2) is the pillar of **Respect for People**. This means empowering those who do the work to improve the process. In a Lean environment, leadership provides the "What" (the challenge), and the team provides the "How" (the solution).

3.4. Summary: From Lean to Agile

Lean provides the **efficiency** and **flow** that allow Agile teams to be truly responsive. Without Lean, Agile teams often suffer from "Agile in name only"—where they work in Sprints but are slowed down by massive backlogs, hand-off delays, and high defect rates.

- **PMBOK 7 Alignment:** Lean supports the principles of *Optimize Leadership Responses* and *Navigate Complexity* by focusing on the delivery system.
- **PMBOK 8 Alignment:** Lean is the driver for **Value Stream Orchestration**. It moves the practitioner from "managing a backlog" to "governing the flow of value" across the entire organizational landscape, ensuring sustainability and high-quality outcomes.

Lean Thinking | The Foundation

Identity

Lean Thinking, 1950s (TPS), Foundational Philosophy.

Core Intent

To maximize customer value and relentlessly eliminate waste.

Operational Heart

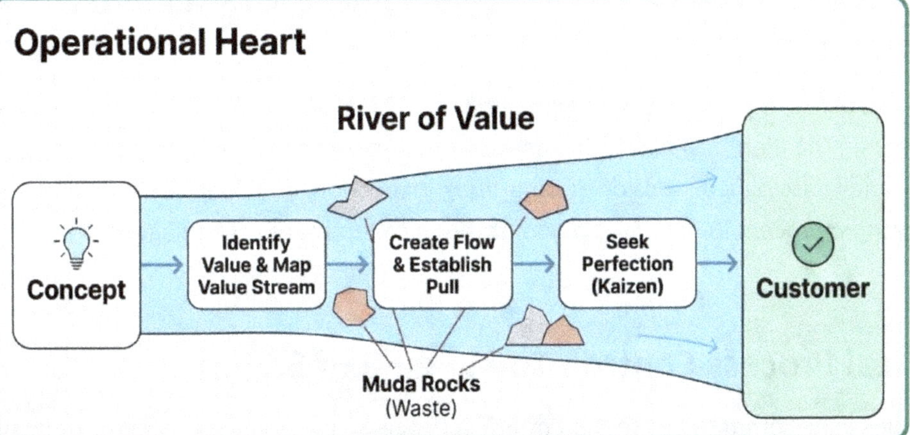

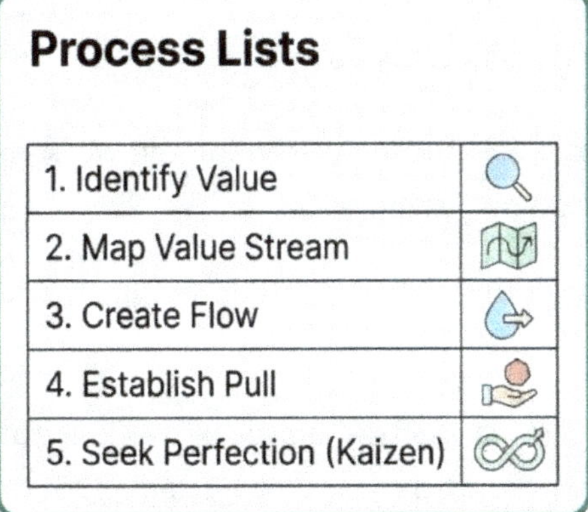

Process Lists

1. Identify Value	
2. Map Value Stream	
3. Create Flow	
4. Establish Pull	
5. Seek Perfection (Kaizen)	

Role Matrix

No prescribed roles; emphasizes "Acts of Leadership" at all levels.

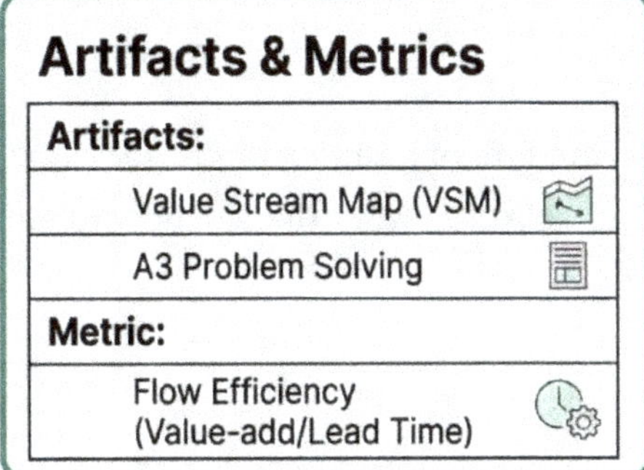

Artifacts & Metrics

Artifacts:	
Value Stream Map (VSM)	
A3 Problem Solving	
Metric:	
Flow Efficiency (Value-add/Lead Time)	

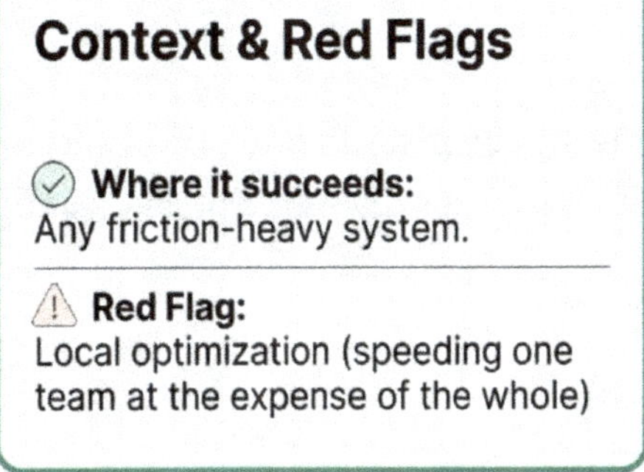

Context & Red Flags

Where it succeeds:
Any friction-heavy system.

Red Flag:
Local optimization (speeding one team at the expense of the whole)

Figure 3-2: Lean Thinking (The Foundation) Snapshot

Chapter 4: Scrum — The Iterative Delivery Framework

4.1. Origin and Ideology

Scrum is the most widely adopted framework in the Agile landscape, but its roots lie not in software, but in product development theory.

- **The Origin:** The term "Scrum" was first introduced by Hirotaka Takeuchi and Ikujiro Nonaka (1986) in their landmark *Harvard Business Review* article, "The New New Product Development Game." They compared high-performing, cross-functional teams to a rugby team moving the ball as a single unit. In the early 1990s, Ken Schwaber and Jeff Sutherland (2020) formalized these concepts into the Scrum framework we use today.
- **The Ideology:** Scrum is built on **Empiricism** and **Lean Thinking**. Its core ideology is that in complex environments, "The Plan" is a fallacy. Instead, Scrum posits that the only way to navigate uncertainty is through small, disciplined loops of transparency, inspection, and adaptation. It shifts the focus from *following a process* to *achieving a goal* through collective team intelligence.

4.2. Empirical Process Control: The Engine of Scrum

Scrum's power lies in its commitment to the **Empirical Process**. In a predictive model, we assume we can plan the entire project at the start. In Scrum, we assume we don't know everything yet.

Every Scrum event is a formal opportunity to implement the three pillars of empiricism:

- **Transparency:** The process and the work must be visible to those performing the work and those receiving it. Without transparency, decisions are based on flawed data.
- **Inspection:** The artifacts and the progress toward goals must be inspected frequently and diligently to detect potential problems or variances.
- **Adaptation:** If an inspector determines that one or more aspects of a process deviate outside acceptable limits, the process or the materials being produced must be adjusted as soon as possible to minimize further deviation.

In practice, teams may be pressured to skip events or rush increments. Understanding Scrum's ideology of small, disciplined loops helps teams maintain focus on value delivery rather than just completing tasks.

4.3. Accountabilities, Events, and Artifacts

Having established Scrum's empirical foundation and ideology, we now examine the concrete roles, events, and artifacts that implement this approach in practice.

Scrum is deliberately incomplete. It provides the "skeletal" structure (The 3-5-3: three accountabilities, five events, and three artifacts) and trusts the team to flesh it out based on their context.

4.3.1. The Scrum Team (Accountabilities)

Scrum moves away from traditional hierarchical silos. The Scrum Team is a cohesive unit of professionals focused on one objective at a time: the Product Goal. There are no sub-teams (e.g., no "testing team") or hierarchies within it. The Scrum Guide (Schwaber & Sutherland, 2020) replaces the concept of "Roles" with Accountabilities to emphasize that while individuals may have different job titles, they share responsibility for value delivery.

- **The Product Owner:** Accountable for **maximizing the value of the product** resulting from the work of the Scrum Team. They are also accountable for effective Product Backlog management, which includes explicitly communicating the **Product Goal**. To succeed, the entire organization must respect their decisions.
- **The Scrum Master:** Accountable for the **Scrum Team's effectiveness**. They do this by enabling the team to improve its practices within the Scrum framework. Unlike older definitions which called them "servant leaders," the modern guide defines them as **"true leaders who serve"** the Scrum Team and the larger organization. This distinction empowers them to take action to remove impediments rather than just facilitating meetings.
- **The Developers:** The people in the Scrum Team that are committed to creating any aspect of a usable Increment each Sprint. They are **Self-Managing**, meaning they internally decide **who** does the work, **when**, and **how**. They are accountable for creating a plan for the Sprint (the Sprint Backlog), instilling quality by adhering to a **Definition of Done**, and adapting their plan each day toward the **Sprint Goal**.

4.3.2. The Events (The Pulse of Scrum)

Scrum events are designed to enable transparency and inspection. All events are timeboxed.

- **The Sprint:** A fixed-length event of one month or less to create consistency. It is the container for all other events.
- **Sprint Planning:** This event initiates the Sprint by laying out the work to be performed. It addresses three specific topics:
    1. **Why is this Sprint valuable?** (The Sprint Goal)
    2. **What can be Done this Sprint?** (Selecting items from the Product Backlog)
    3. **How will the chosen work get done?** (The plan of action)

- **Daily Scrum:** A 15-minute event for the **Developers** to inspect progress toward the Sprint Goal and adapt the Sprint Backlog as necessary. The specific "three questions" format is no longer prescribed; developers can conduct this meeting however they see fit, provided it focuses on progress toward the **Sprint Goal** and produces an actionable plan for the next day.
- **Sprint Review:** An informal session to inspect the outcome of the Sprint and determine future adaptations. The Scrum Team presents the results of their work to key stakeholders, and progress toward the Product Goal is discussed.
- **Sprint Retrospective:** A session to plan ways to increase quality and effectiveness. The team inspects how the last Sprint went with regards to individuals, interactions, processes, tools, and their Definition of Done.

4.3.3. The Artifacts and Commitments

To ensure transparency and focus, each of the three Scrum Artifacts is inextricably linked to a specific **Commitment**. These commitments serve as the standard against which progress is measured:

- **Product Backlog (Commitment: Product Goal):** An emergent, ordered list of what is needed to improve the product. The **Product Goal** describes the future state of the product which can serve as a target for the Scrum Team to plan against.
- **Sprint Backlog (Commitment: Sprint Goal):** The set of Product Backlog items selected for the Sprint, plus a plan for delivering them. The **Sprint Goal** is the single objective for the Sprint, creating coherence and focus rather than separate initiatives.
- **Increment (Commitment: Definition of Done):** A concrete stepping stone toward the Product Goal. The **Definition of Done** is a formal description of the state of the Increment when it meets the quality measures required for the product. If a backlog item does not meet the Definition of Done, it is not an Increment and cannot be released.

4.3.4. The Definition of Outcome Done (DoOD)

While the traditional Definition of Done (DoD) focuses on the quality of the *output* (the code/feature), modern Scrum in 2026 distinguishes this from the **Definition of Outcome Done (DoOD)**. Introduced in the *Scrum Guide Expansion Pack* (Jocham et al., 2025), the DoOD asks: "Did this increment actually change customer behavior?"

- **DoD (Output):** The code is written, tested, and deployed.
- **DoOD (Outcome):** The customer has used the feature, and we have measured a change in their behavior (e.g., retention increased by 2%).

This shift forces the Product Owner to prioritize validation over mere delivery, moving the team from a "Feature Factory" to a "Value Engine".

4.4. The Definition of Done (DoD) vs. Acceptance Criteria

To ensure transparency, Scrum requires a clear standard for quality. It is vital to distinguish between these two concepts:

- **Acceptance Criteria:** These are unique to each individual User Story or Product Backlog Item. They define the specific functional requirements that a specific feature must satisfy to be accepted by the Product Owner.
- **Definition of Done (DoD):** This is a formal, shared description of the state of the Increment when it meets the quality measures required for the product. It applies to **all** items in the Sprint. If an item does not meet the DoD, it cannot be released or even presented at the Sprint Review.

Analogy: If you are building a house, the **Acceptance Criteria** might state that the kitchen must have granite countertops. The **Definition of Done** states that the electricity must be grounded and the plumbing must not leak—regardless of which room you are building.

4.5. Strengths, Limitations, and When to Use Scrum

Scrum excels in environments where the "What" and "How" are uncertain. However, it requires a high degree of trust and organizational support. It is less effective in environments where work is highly repetitive or where external dependencies make fixed timeboxes impossible to maintain.

4.6. Framework Operational Profile: Scrum

To provide a concise reference and enable comparison with other frameworks, Scrum's practices are summarized in the Operational Profile below.

Table 4-1: Framework Operational Profile: Scrum

Feature	Description
1. Agile Approach Category	**Iterative / Timeboxed**
2. Core Intent / Purpose	To solve complex problems and deliver high value through adaptive solutions.
3. Primary Focus	**Value Delivery & Adaptability (via Empirical Process Control).**
4. Roles (Accountabilities)	**Product Owner** (Value), **Scrum Master** (Effectiveness/True Leader), **Developers** (Creation/Quality).
5. Cadence / Timing Model	Fixed iterations (**Sprints**) of 1 month or less.
6. Core Practices / Ceremonies	Sprint Planning, Daily Scrum, Sprint Review, Sprint Retrospective.
7. Key Artifacts & Commitments	Product Backlog (**Product Goal**), Sprint Backlog (**Sprint Goal**), Increment (**DoD/DoOD**).
8. Work Management Model	**Self-Managing**: Team pulls work and decides "how," "who," and "when."
9. Primary Metrics[1]	Value Delivered, Cycle Time, Sprint Goal Achievement. *(Note: Velocity is a complementary practice, not a core Scrum rule).*
10. Organizational Fit	Small cross-functional teams (typically 10 or fewer); complex product environments.
11. Strengths	Rapid risk mitigation; high transparency; fosters strong team ownership and alignment to business goals.
12. Common Failure Modes	**"Zombie Scrum"** (ritual without mindset); **"Scrum-but"** (omitting key elements); Lack of psychological safety; PO lacking authority.
13. Best Used When...	The problem is complex (unknown requirements/technology) and requires frequent feedback loops to validate assumptions.

[1] These metrics are intended to guide team learning and transparency rather than to punish or rate individual performance.

Scrum | Iterative Delivery

Identity

Scrum, 1995, Iterative/Timeboxed Framework.

Core Intent

To navigate complex problems through small loops of empirical learning.

Operational Heart

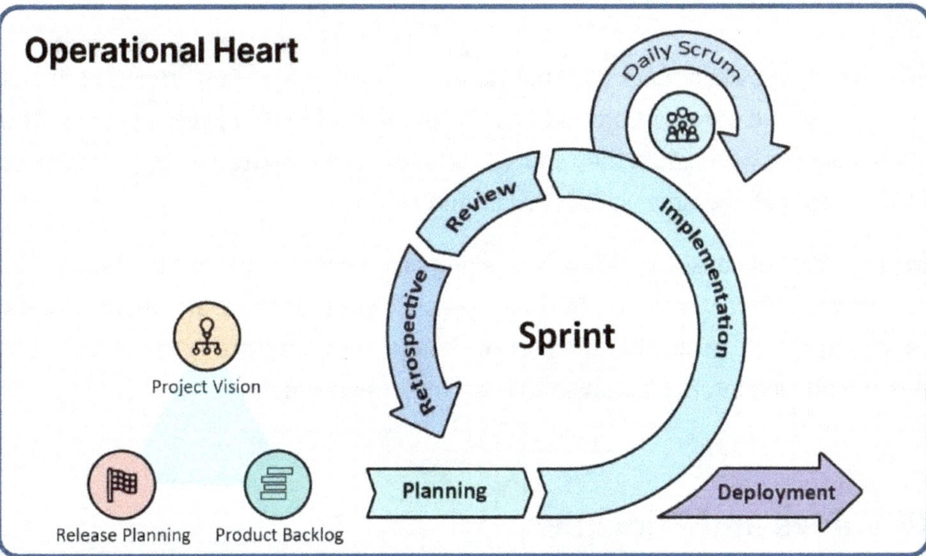

Process Lists

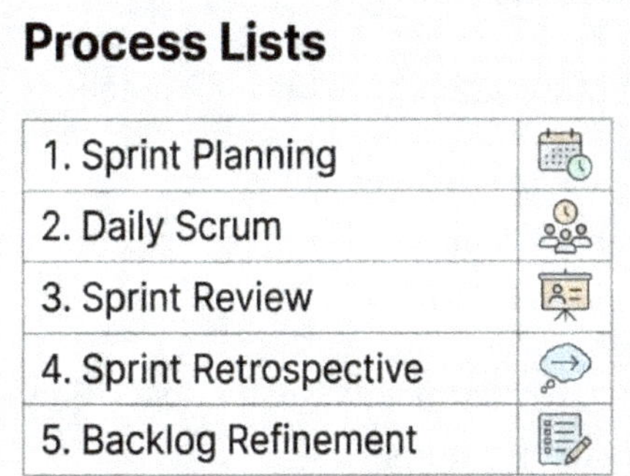

1. Sprint Planning
2. Daily Scrum
3. Sprint Review
4. Sprint Retrospective
5. Backlog Refinement

Role Matrix

Artifacts & Metrics

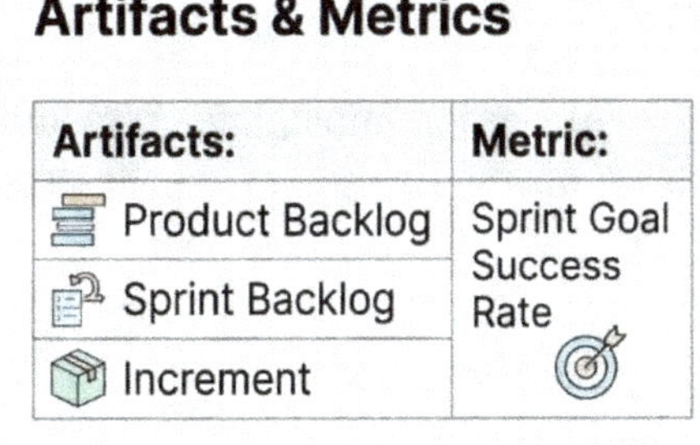

Artifacts:	Metric:
Product Backlog	Sprint Goal Success Rate
Sprint Backlog	
Increment	

Context & Red Flags

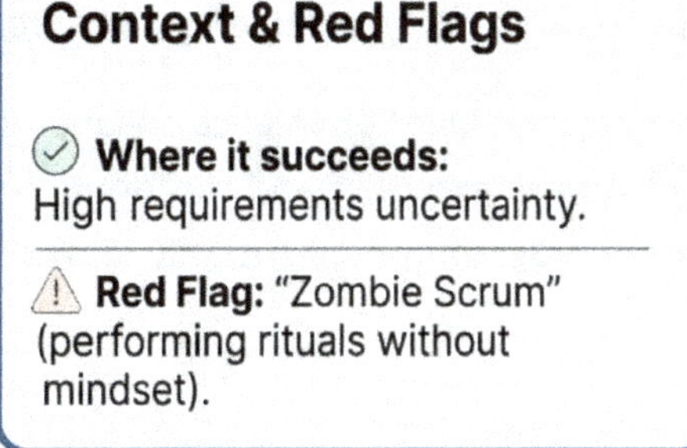

Figure 4-1: Scrum (Iterative Delivery) Framework Snapshot

Chapter 5: Extreme Programming (XP) — The Engineering Heart of Agile

5.1. Origin and Ideology

While many Agile frameworks focus on project management, XP was born from the need to improve software quality and responsiveness to changing customer needs. XP's ideology reinforces the Agile principle of delivering high-value increments by ensuring that each Increment is technically robust and adaptable to change.

- **The Origin:** Developed by Kent Beck (2004), Ward Cunningham, and Ron Jeffries in the late 1990s during their work on the Chrysler Comprehensive Compensation (C3) payroll system. Beck realized that if certain practices were good, they should be taken to the "extreme" (e.g., if code review is good, we should do it all the time via Pair Programming).
- **The Ideology: XP** delivers business agility by making **technical excellence** the engine of empirical feedback. Its ideology is centered on five core values: **Communication, Simplicity, Feedback, Courage, and Respect**. XP posits that the cost of change does not have to rise exponentially over time if we maintain high internal quality and keep the design simple.

5.2. The XP Values and Principles

XP is unique because it explicitly links human values to specific engineering practices.

- **Communication:** Problems in projects are often communication problems. XP practices are designed to force frequent, face-to-face interaction.
- **Simplicity:** We will do what is needed and asked for, but no more. This maximizes the amount of work *not* done (the "YAGNI" principle: *You Ain't Gonna Need It*).
- **Feedback:** By demonstrating software early and often, the team gets the feedback they need to stay on track. This applies to the code (tests) and the product (customer).
- **Courage:** The courage to tell the truth about progress, to throw away code that isn't working, and to stand by the team's quality standards under pressure.
- **Respect:** Because XP relies on intense collaboration (like pair programming), team members must respect each other's expertise and contributions. Respect for the customer means delivering value; respect for the programmer means a sustainable pace.

5.3. Core XP Practices

XP is often visualized as a set of interlocking circles (the "XP Circle of Life"), ranging from fine-scale feedback to outer-loop project management. These practices operationalize XP values and create the feedback loops needed for empirical process control.

5.3.1. Fine-Scale Feedback (The Inner Loop)

- **Pair Programming:** Two developers at one workstation. This ensures continuous code review and knowledge sharing.
- **Test-Driven Development (TDD):** Writing a failing automated test before writing the code itself. This ensures the code is testable and meets requirements from the start.
- **Ten-Minute Build:** The entire system should be able to be built and all tests passed in ten minutes or less to ensure rapid feedback.

5.3.2. Continuous Process

- **Continuous Integration (CI):** Developers integrate their code into the main branch several times a day.
- **Refactoring:** Constantly improving the internal structure of the code without changing its external behavior to keep the design "clean."
- **Small Releases:** Putting a functional system into production quickly and updating it frequently.

5.3.3. Shared Understanding

- **Simple Design:** The right design for the software is the simplest one that works for the current requirements.
- **System Metaphor:** A shared story or analogy that describes how the whole system works, ensuring everyone uses the same vocabulary.
- **Collective Code Ownership:** Anyone on the team can improve any part of the code at any time.

These practices enable empiricism by making progress visible (Transparency), outcomes inspectable (Inspection), and allowing rapid adaptation.

5.4. Technical Debt and the "Flaccid Scrum" Trap

A critical concept in the Agile landscape is **Technical Debt**. This occurs when teams prioritize speed over quality, leading to "messy" code that is hard to change. Without XP practices, Scrum teams risk delivering fragile software despite following Agile rituals.

In many organizations, teams adopt **Scrum** (the management rituals) but ignore **XP** (the technical practices). This often leads to "Flaccid Scrum"—where the team looks Agile on the outside (they have Sprints and Stand-ups), but they cannot actually deliver a "Done" increment because their codebase is too fragile to change. XP is the antidote to this trap.

For example, a team might finish two-week Sprints, but if automated tests are skipped, adding a new feature could break unrelated functionality—highlighting why XP practices are essential.

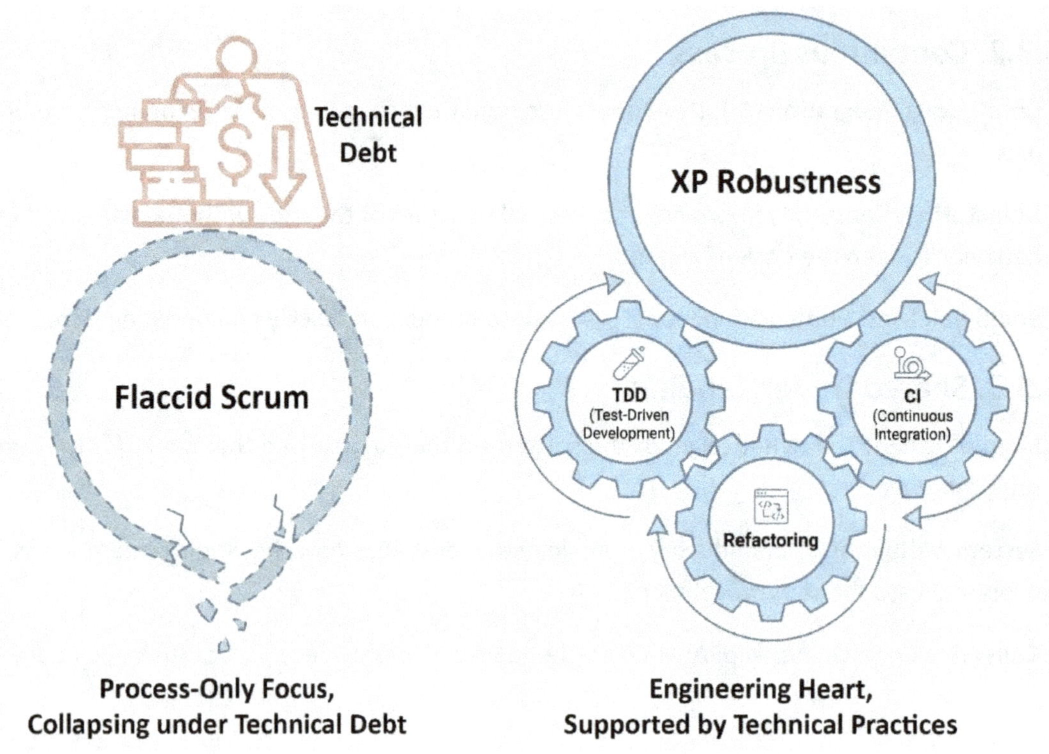

Figure 5-1: The "Flaccid Scrum" Trap vs. XP Robustness

5.5. Framework Operational Profile: Extreme Programming (XP)

To provide a concise reference and enable comparison with other frameworks, XP's practices are summarized in the Operational Profile below.[1]

Table 5-1: Framework Operational Profile: XP

Feature	Description
1. Agile Approach Category	Engineering-Focused / Iterative
2. Core Intent / Purpose	To improve software quality and responsiveness through extreme engineering discipline.
3. Primary Focus	**Technical Excellence & Quality:** Reducing the cost of change through continuous feedback and simple design.
4. Roles	Customer (defines value), Programmer (delivers code), Coach (mentor), Tracker (metrics).
5. Cadence / Timing Model	Short iterations (often 1 week); very frequent releases (daily or weekly).
6. Core Practices / Ceremonies	Pair Programming, TDD, CI, Refactoring, Co-located team, Whole Team participation.
7. Key Artifacts	User Stories (on cards), Automated Test Suite, Clean Code.
8. Work Management Model	**Values-based:** Driven by the Customer's "Planning Game" and Programmer estimates.
9. Primary Metrics	Velocity, Build Time, Test Coverage, Defect Count
10. Organizational Fit	Small to medium teams (2-12 people); software-intensive projects with high technical complexity; quality is critical.
11. Strengths	Extremely high code quality; very low defect rates; high team morale and knowledge sharing.
12. Common Failure Modes	Cultural resistance to Pair Programming; skipping TDD under pressure; "Practice-only" adoption without the Values.
13. Best Used When...	Technical risks are high, the cost of failure is significant, or the project requires a long-term sustainable pace.

[1] XP's focus on technical excellence differentiates it from other iterative frameworks.

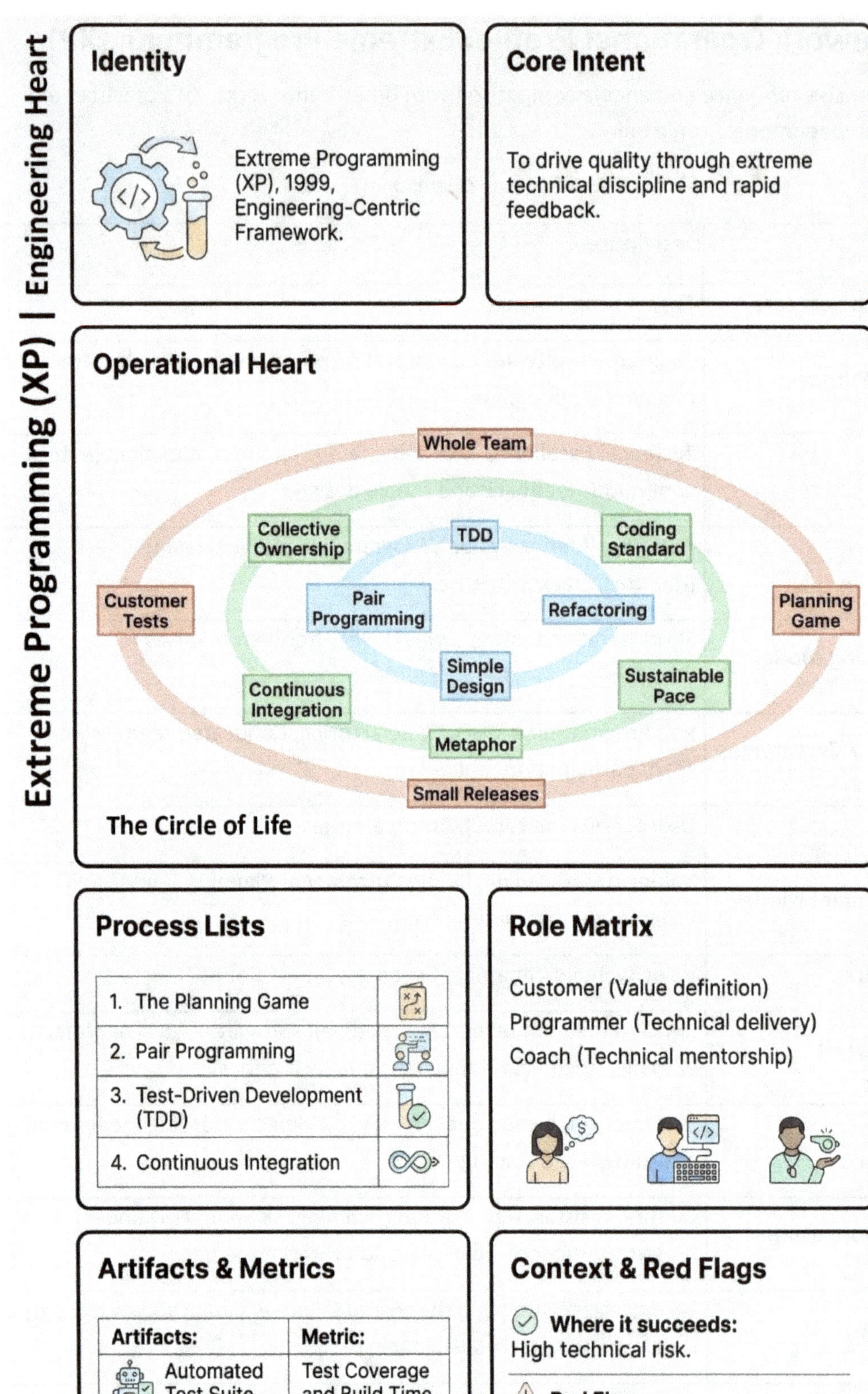

Figure 5-2: Extreme Programming (Engineering Heart) Framework Snapshot

Chapter 6: Kanban — The Flow-Based Landscape

6.1. Origin and Ideology

Kanban is more than a visual board; it is a strategy to optimize flow of value. Kanban optimizes flow by making work visible and limiting work in progress, without requiring structural changes to roles or meetings.

- **The Origin:** "Kanban" (Japanese for "signboard" or "visual signal") originated in the **Toyota Production System** as a way to manage inventory. In the early 2000s, David J. Anderson (2010) adapted these concepts for knowledge work and software development, creating the "Kanban Method."
- **The Ideology:** Unlike Scrum, which requires a significant structural change (new roles, new meetings), Kanban is an **evolutionary** framework. Its core ideology is: *"Start with what you do now."* It does not change your roles or process overnight; instead, it makes the existing process visible so that it can be improved incrementally. It shifts the focus from managing people to **managing the work**.

6.2. The Core Principles and General Practices

The Kanban Method is defined by two sets of principles (Change Management and Service Delivery) and six general practices.

6.2.1. Change Management Principles

- **Start with what you do now:** Understand current processes and respect existing roles and titles.
- **Agree to pursue evolutionary change:** Improvement should be continuous and incremental, not radical and disruptive.
- **Encourage acts of leadership at all levels:** From individual contributors to senior management.

6.2.2. The Six General Practices

- **Visualize the Work:** Create a Kanban Board that shows the flow of work from "To Do" to "Done."
- **Limit Work in Progress (WIP):** Restrict the number of items in any stage to improve flow and reduce lead time.
- **Manage Flow:** Observe how work moves through the system and address bottlenecks.
- **Make Process Policies Explicit:** Clearly define how work is handled (e.g., "Definition of Ready").
- **Implement Feedback Loops:** Use cadences like the "Kanban Meeting" (daily) or "Operations Review."
- **Improve Collaboratively, Evolve Experimentally:** Use models and the scientific method to drive change.

Agile in Hard Hats:

The Last Planner System (LPS) If you come from the construction world, Kanban will look familiar. It is the spiritual sibling of the **Last Planner System (LPS)** used in Lean Construction. Both systems shift from "Push" (following a master schedule) to "Pull" (doing work only when constraints are removed).

- **Phase Scheduling** = Release Planning.
- **Make-Ready Planning** = Backlog Refinement.
- **Weekly Work Plan** = Sprint Backlog.
- **Percent Plan Complete (PPC)** = The reliability metric, similar to "Say/Do Ratio" in Scrum.

Just as Kanban limits WIP to ensure flow, LPS ensures that no work is promised unless it can actually be performed, preventing the "stop-start" chaos of traditional job sites.

6.3. The Power of WIP Limits and Little's Law

The defining characteristic of Kanban is the move from a "Push" system to a **"Pull" system**.

In a push system (like traditional Waterfall), work is assigned to people regardless of their current load. In Kanban, a new task can only be "pulled" into a column if there is capacity available under the **WIP Limit**.

This is based on **Little's Law**, a mathematical principle applied to product development flow by Donald Reinertsen (2009), which states that the more work you have in the system, the longer each piece of work takes to finish. By limiting WIP, Kanban enables rapid feedback on system performance, improving throughput and reducing lead time.

6.4. Kanban vs. Scrum: The "Context-Driven" Choice

In *The Agile Landscape*, Kanban and Scrum are complementary; choose based on work type and team needs:

- **Choose Scrum if:** You have a clear product goal, a dedicated team, and the work can be planned in 1–4 week chunks.
- **Choose Kanban if:** You have a high degree of unplanned work (support, maintenance), you need to change priorities at any moment, or you are an established team looking to optimize an existing process without changing roles.

6.5. Framework Operational Profile: Kanban

To provide a concise reference and enable comparison with other frameworks, Kanban's practices are summarized in the Operational Profile below.[1]

Table 6-1: Framework Operational Profile: Kanban

Feature	Description
1. Agile Approach Category	**Flow-Based / Continuous**
2. Core Intent / Purpose	To visualize work, limit WIP, and optimize the flow of value through a system.
3. Primary Focus	**Flow & Throughput:** Reducing Lead Time (the time it takes to go from start to finish).
4. Roles	No specific roles required (evolutionary); often uses "Service Delivery Manager" or "Service Request Manager."
5. Cadence / Timing Model	**Continuous Flow:** No fixed iterations; events (like replenishment) happen on an as-needed or fixed cadence.
6. Core Practices / Ceremonies	Kanban Meeting (Daily), Replenishment Meeting, Delivery Planning, Service Delivery Review.
7. Key Artifacts	Kanban Board, WIP Limits, Cumulative Flow Diagram (CFD).
8. Work Management Model	**Pull-based:** Capacity-limited system where work is pulled when a "slot" becomes available.
9. Primary Metrics	Cycle Time, Lead Time, Throughput, WIP, Service Level Expectation (SLE).
10. Organizational Fit	Operations, Support, Marketing, or highly mature dev teams; works for teams of any size.
11. Strengths	Flexibility; reduces multitasking; low resistance to adoption; highly data-driven.
12. Common Failure Modes	**"The Visual-Only Board"** (ignoring WIP limits); lack of discipline in updating the board; local optimization.
13. Best Used When...	Work is continuous/variable; speed of delivery is more important than iteration-based predictability.

[1] Kanban emphasizes flow optimization over iteration, making it suitable for continuous or variable work.

Kanban | Flow-Based Landscape

Identity

Kanban (Flow-Based Landscape), 2010 (Knowledge Work)

Evolutionary Flow Framework

Core Intent

To visualize work and limit Work-in-Progress (WIP) to optimize throughput.

Operational Heart

To Do	Doing	Done
	WIP Limit:3	
Task 6 →	Task 3 →	Task 1
Task 7 →	Task 4 →	Task 2
Task 8 →	Task 5 →	

Pull signal triggered

Process Lists

Main processes:

1. Replenishment
2. Kanban Meeting (Daily)
3. Delivery Planning
4. Service Delivery Review

Role Matrix

- Evolutionary (starts with existing roles); often uses Service Delivery Manager (Flow).

Artifacts & Metrics

Artifacts:
Kanban Board
Cumulative Flow Diagram (CFD)
Metric:
Lead Time (CycleTime=WIP/Throughput)

Context & Red Flags

Where it succeeds:
Unplanned/Continuous work.

Red Flag:
Visual-only board (no WIP limits).

Figure 6-1: Kanban (Flow-Based Landscape) Framework Snapshot

Part III: The Agile Family and Hybrid Approaches

Chapter 7: The Landscape of Choice

7.1. Context, Constraints, and Trade-Offs

In the pursuit of organizational agility, the most dangerous pitfall is methodology dogma. To navigate the landscape effectively, a practitioner must internalize a fundamental truth: **"No framework is universally correct."** The success of an approach is not found in the framework itself, but in the tight alignment between that framework and the specific context of the work.

Choosing a way of working (WoW) requires a cold-eyed assessment of five critical decision dimensions:

1. **Predictability vs. Uncertainty:** Does the project require a high degree of upfront certainty (e.g., fixed-price regulatory work), or is it an exploratory venture where the requirements emerge only through delivery?
2. **Flow vs. Cadence:** Is the work better served by a "heartbeat" of regular timeboxes (Cadence), or does it require a continuous stream of delivery where priorities shift by the hour (Flow)?
3. **Engineering Intensity:** Does the product have high technical risk or safety-critical requirements that demand extreme rigor, or is it a low-risk internal tool?
4. **Team Maturity:** Is the team composed of seasoned Agile practitioners capable of self-management, or do they require more prescriptive guardrails as they build their foundational skills?
5. **Organizational Constraints:** What are the non-negotiables? These include geographical distribution, existing governance structures, and the culture's tolerance for radical transparency.

7.2. Choosing the Right Method for the Right Problem

Strategic selection is the process of matching these dimensions to the strengths of the **15 delivery and scaling frameworks** explored in this guide. While Lean Thinking (Chapter 3) provides the foundational mindset and efficiency principles required to make these choices, the specific execution models used for comparative selection are detailed in the subsequent chapters. The goal is to move from "doing a framework" to "solving a problem."

- **The Scrum Choice:** Often effective when uncertainty is high but the team can commit to a stable cadence. It provides the necessary structure to build a product vision through iterative feedback.
- **The Kanban Choice:** Best used when the primary constraint is lead time and the work is highly variable. It excels in environments where "stop starting, start finishing" is more valuable than long-range planning.
- **The XP Choice:** Necessary when the engineering intensity is high. Even if using Scrum for management, XP's technical practices are often the "muscle" required to keep the landscape adaptable.
- **The DSDM/FDD Choice:** Effective when organizational constraints demand higher governance or when the project is architecture-heavy and requires model-driven stability.

Several of these approaches—particularly those shaped by governance or architectural constraints—will be examined in greater depth in the historical frameworks presented in Chapters 8–11. Ultimately, the "Landscape of Choice" is a matter of professional judgment. Experienced teams rarely adopt a framework wholesale; instead, they deliberately design a Way of Working that combines practices to balance flexibility, predictability, and real-world constraints.

Chapter 8: Feature-Driven Development (FDD)

8.1. Origin and Ideology

While frameworks like Scrum were emerging in the U.S., FDD was developed to handle the complexities of large-scale, high-stakes banking software in Southeast Asia.

- **The Origin:** FDD was devised by **Jeff De Luca** and **Peter Coad** (Palmer & Felsing, 2002) in the late 1990s during a massive 15-month project for a large bank in Singapore. Unlike other Agile methods that started with small teams, FDD was "born large," designed from day one to coordinate dozens of developers.
- **The Ideology:** The core philosophy of FDD is that **quality is a byproduct of sound architecture and frequent, tangible results.** While it remains firmly Agile—delivering working software in short iterations—it is more structured and "top-down" than Scrum. It values the role of the "Chief Architect" and "Class Owners," believing that a shared, high-level model is the best way to prevent a large system from fragmenting into technical chaos.

8.2. Model-Driven and Feature-Centric Delivery

FDD is highly structured, revolving around a five-step process. This process ensures that every piece of code written is directly linked to a business feature and a global domain model.

8.2.1. The Five-Step Process

1. **Develop an Overall Model:** The project begins with high-level walkthroughs of the system's scope. Designers and the Chief Architect create a "shape" for the system.
2. **Build a Feature List:** The team identifies features—small, client-valued functions (expressed as "verb-result-object," e.g., "Calculate the total of a sale").
3. **Plan by Feature:** Features are sequenced into "Feature Sets" and assigned to programmers.
4. **Design by Feature:** A small group of developers designs the specific features for a 1–2 week work package.
5. **Build by Feature:** The code is developed, tested, and integrated.

8.2.2. The Critical Strategic Choice: Individual vs. Collective Ownership

A unique aspect of FDD is **Individual Class Ownership**. Unlike the "Collective Code Ownership" of XP (where any developer can change any code), FDD assigns specific pieces of code (classes) to specific developers.

This contrast is a perfect example of **"Context-Driven" selection**:

- **Choose XP (Collective Ownership)** when you prioritize speed, resilience, and removing "key person dependencies" (the Bus Factor).
- **Choose FDD (Individual Ownership)** when you prioritize architectural rigidity and individual accountability. This is often required in **low-trust** or **high-compliance** environments where it must be explicitly clear *who* changed a critical algorithm and why.

8.3. Why FDD Worked— and Its Modern Niche

8.3.1. Why FDD Worked

In the modern landscape, FDD is often overlooked because it feels more "formal" than the popular iterative frameworks. However, this formality is its greatest strength in specific contexts:

- **Large-Scale Engineering:** Because FDD relies on a central model and Chief Architects, it scales more naturally than "Scrum of Scrums" for very large teams (50+ people) working on a single, complex codebase.
- **Reporting and Transparency:** FDD's granular feature list allows for very precise progress tracking. Managers can see exactly what percentage of "Feature Sets" are complete, making it highly effective for executive reporting in traditional corporate environments.

8.3.2. Why It Faded

Despite its strengths, FDD gradually fell out of favor as Agile adoption moved toward smaller, cross-functional teams and lighter governance models. Its reliance on centralized architectural authority and individual code ownership increasingly conflicted with the industry's shift toward DevOps and continuous delivery, where silos are discouraged.

8.3.3. What Survived: The "Legacy" Bridge

While FDD acts as a "legacy" framework, it remains a vital historical bridge. It demonstrates how the heavy governance of traditional **Rational Unified Process (RUP)** was distilled into Agile practices. For the predictive project manager, FDD offers a comfortable middle ground: it provides the predictability of a model-driven approach while maintaining the speed of 2-week feature cycles.

8.4. Framework Operational Profile

To provide a concise reference and enable comparison with other frameworks, FDD's practices are summarized in the Operational Profile below.

Table 8-1: Framework Operational Profile: FDD

Feature	Description
1. Agile Approach Category	**Architecture-Focused / Feature-Centric**
2. Core Intent / Purpose	To deliver functional, high-quality software through disciplined modeling and short feature-build cycles.
3. Primary Focus	**Technical Discipline & Scalability:** Managing complexity through a shared domain model and clear ownership.
4. Roles	Chief Architect, Project Manager, Development Manager, Class Owners (Developers), Domain Experts.
5. Cadence / Timing Model	**Feature-based:** Short cycles (typically 2–14 days) per feature.
6. Core Practices / Ceremonies	Domain Modeling, Developing by Feature, Individual Class Ownership, Regular Build.
7. Key Artifacts	Domain Model, Feature List (Work Packages), Design Packages.
8. Work Management Model	**Plan-by-Feature:** Highly organized list-driven development; features are assigned to Class Owners.
9. Primary Metrics	**Feature Progress (Milestones)**, **Build Health**, **Code Quality per Class**.
10. Organizational Fit	Large teams; complex, architecture-heavy systems; High-Compliance environments.
11. Strengths	Scalability; high architectural integrity; clear individual accountability; excellent for reporting.
12. Common Failure Modes	**Bottlenecks:** If a Class Owner is unavailable, work on that component stops; potentially slower "start-up" due to initial modeling phase.
13. Best Used When...	Building complex systems where individual accountability is required (e.g., banking, defense), or trust levels are too low for Collective Ownership.

Feature-Driven Development | FDD

Identity

Feature-Driven Development (FDD), 2002, Architecture-Focused scaling.

Core Intent

To deliver quality results through model-driven feature cycles.

Operational Heart

Model (Develop Overall Model)	Feature List (Build Feature List)	Plan (Plan by Feature)	Design (Design by Feature)	Build (Build by Feature)
Conceptual model	Single-weight structure	Calendar flow diagram	Technical drawign	Construction flow diagram

Process Lists

1. Domain Modeling
2. Designing by Feature
3. Building by Feature
4. Code Inspections

Role Matrix

Chief Architect

Class Owners (Individual Ownership)

Domain Experts

Artifacts & Metrics

Artifacts:
Feature List (Verb-Result-Object)
Design Packages
Metric:
Feature Progress (%) 58%

Context & Red Flags

Where it succeeds:
Very large teams on complex OO systems.

Red Flag:
Siloing and lack of collective code ownership.

Figure 8-1: Feature-Driven Development (FDD) Framework Snapshot

Chapter 9: Dynamic Systems Development Method (DSDM)

9.1. Origin and Ideology

DSDM was the answer to a specific corporate problem: How can we reap the benefits of rapid development without losing control of the project?

- **The Origin:** Established in 1994 by a non-profit consortium, the framework was formalized as the **DSDM Agile Project Framework** (Agile Business Consortium, 2014) to provide a standardized, disciplined framework for **Rapid Application Development (RAD)**. It embodied many Agile principles before the term "Agile" was formally coined for software development, and its creators were among the signatories of the Agile Manifesto.
- **The Ideology:** DSDM is built on the philosophy that "any project must be aligned to clearly defined strategic goals and focus on early delivery of real benefits to the business." Unlike frameworks that focus on the team's internal mechanics, DSDM focuses on the **entire project lifecycle**, providing a bridge between the flexible world of development and the structured world of corporate governance.

9.2. Governance-Friendly Agile: The Inverted Triangle

One of DSDM's most significant contributions to the Agile landscape is the **Inverted Triangle** model of project constraints.

In traditional project management (Waterfall), the **Scope** of a project is fixed, while **Time** and **Cost** are variable (leading to frequent delays and budget overruns). DSDM flips this on its head:

- **Fixed:** Time, Cost, and Quality. The deadline and budget are non-negotiable.
- **Variable:** Features (Scope).

This model is "governance-friendly" because it guarantees that something will be delivered on a specific date for a specific price. To manage this variable scope, DSDM utilizes the **MoSCoW prioritization technique** (Agile Business Consortium, 2014) (Must have, Should have, Could have, Won't have this time). This ensures that the business always receives the most critical value—the "Must Haves" (roughly the top 60% of priority items in most cases)—even if lower-priority features are deferred.

9.3. Timeboxing, Roles, and Control

DSDM provides more structure than Scrum, defining a clear set of roles and a lifecycle that covers everything from the Business Case to Post-Project benefits realization.

9.3.1. Roles with Business Clarity

DSDM defines roles that explicitly represent the business interests, ensuring that developers aren't working in a vacuum:

- **Business Sponsor:** The "Owner" of the project who provides the budget and high-level direction.
- **Business Visionary:** Ensures the solution will actually meet the business needs.
- **Solution Development Team:** A cross-functional unit of developers and business ambassadors working in tight synchronization.

9.3.2. Structured Timeboxing

While Scrum has Sprints, DSDM uses **Timeboxes**. A DSDM timebox is more than just a duration; it is a mini-project with three internal phases:

1. **Kick-off:** Confirming the goal and MoSCoW priorities for that window.
2. **Iterative Development:** The actual building and testing.
3. **Close-out:** Formalizing the acceptance and documenting lessons learned.

This structure makes DSDM timeboxes more prescriptive than Scrum Sprints, reinforcing predictability and auditability.

9.3.3. Why It Faded—and What Persisted

DSDM (now rebranded as **AgilePM** by the Agile Business Consortium) is often perceived as "heavy" by modern startups. It requires more documentation and role-definition than a "light" Scrum implementation. However, as organizations grow and face regulatory audits or fixed-price contracts, they often find themselves "reinventing" the practices that DSDM has had in place for decades.

Its emphasis on fixed time and cost, explicit business ownership, and value-based prioritization survives today in enterprise Agile governance, AgilePM, and scaled delivery models.

9.4. Framework Operational Profile

To provide a concise reference and enable comparison with other frameworks, DSDM's practices are summarized in the Operational Profile below.

Table 9-1: Framework Operational Profile: DSDM

Feature	**Description**
1. Agile Approach Category	**Governance-Focused / Project Framework**
2. Core Intent / Purpose	To provide a disciplined agile framework that ensures business value is delivered within fixed time and cost constraints.
3. Primary Focus	Strategic Alignment & Control: Bridging the gap between corporate governance and iterative delivery.
4. Roles	Business Sponsor, Business Visionary, Business Ambassador, Team Leader, Solution Developers.
5. Cadence / Timing Model	**Fixed Timeboxes:** Nested within a larger project lifecycle (Feasibility, Foundations, Evolutionary Development).
6. Core Practices / Ceremonies	MoSCoW Prioritization, Timeboxing, Prototyping, Facilitated Workshops.
7. Key Artifacts	Business Case, Prioritized Requirements List (PRL), Solution Architecture Definition (SAD).
8. Work Management Model	**Fixed Time/Cost, Variable Scope:** Managing delivery through strict prioritization of the backlog.
9. Primary Metrics	**On-time Delivery**, **Budget Adherence**, **Value Realization (ROI)**.
10. Organizational Fit	Large corporations; government/regulated sectors; projects with fixed-price or fixed-date constraints.
11. Strengths	Strong governance; predictable delivery dates; high business engagement; clear accountability.
12. Common Failure Modes	**Over-documentation:** Reverting to Waterfall by being too rigid with "Foundations" documentation.
13. Best Used When...	The project exists within a traditional corporate environment where "Agile" needs to speak the language of "Project Management."

Dynamic Systems Development Method | DSDM

Identity

DSDM, 1994,
Governance-Friendly
framework.

Core Intent

To deliver business value within fixed time and budget constraints.

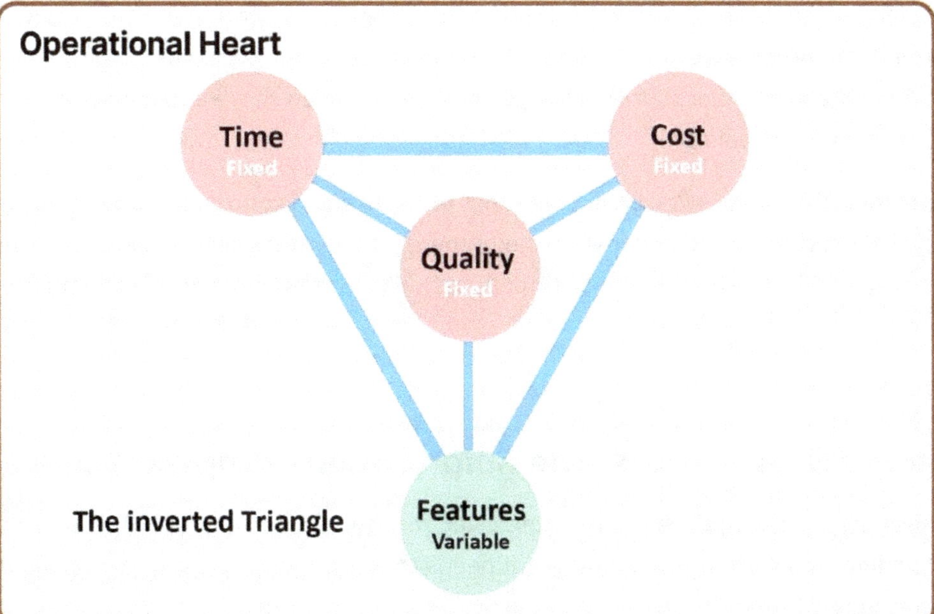

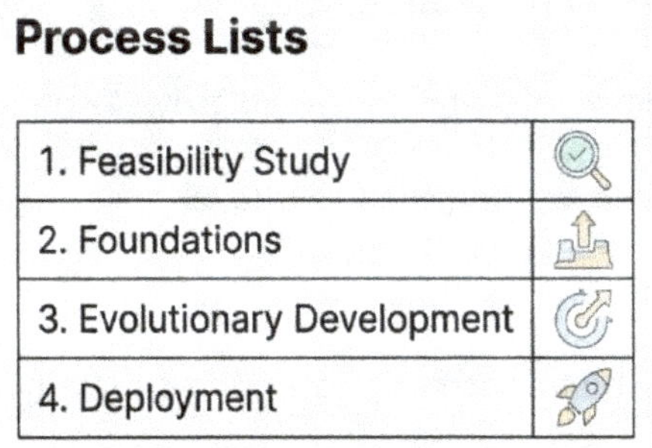

Process Lists

Process
1. Feasibility Study
2. Foundations
3. Evolutionary Development
4. Deployment

Role Matrix

Business Sponsor
Business Visionary
Team Leader
Solution Developer

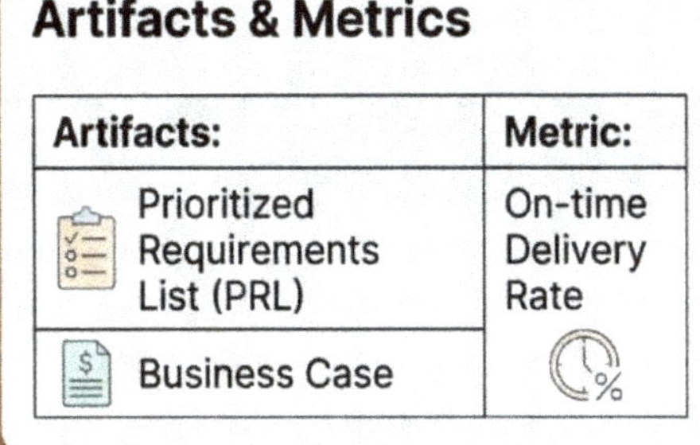

Artifacts & Metrics

Artifacts:	Metric:
Prioritized Requirements List (PRL)	On-time Delivery Rate
Business Case	

Context & Red Flags

Where it succeeds:
Regulated/Corporate environments.

Red Flag:
Over-documentation and rigid Waterfall gates.

Figure 9-1: Dynamic Systems Development Method (DSDM) Framework Snapshot

Chapter 10: Crystal — Human-Centric Agile Methods

10.1. Origin and Ideology

Crystal is not a single framework, but a family of methodologies designed to be "situationally specific."

- **The Origin:** Developed by **Alistair Cockburn** (2004) in the early 1990s, the Crystal family of methodologies was born from his study of successful IBM project teams. He discovered that the most effective teams didn't follow a rigid manual; instead, they communicated frequently and adjusted their "process weight" based on their specific environment. As one of the primary authors of the Agile Manifesto, Cockburn's findings heavily influenced the "People and interactions over processes and tools" value.
- **The Ideology:** Crystal is built on the belief that projects are **"Economic-Cooperative Games."** The goal of the game is to deliver software and prepare for the next game. Because humans are variable and unpredictable, the methodology emphasizes skills, communication, and community over rigid discipline or sub-processes.

10.2. People Over Process: The Color-Coded Complexity Scale

The defining characteristic of Crystal is its categorization system. Cockburn (2004) realized that a three-person team building a marketing site has different needs than a hundred-person team building a nuclear reactor control system.

Crystal uses two dimensions to determine the "weight" of the methodology:

1. **Team Size:** As more people join a project, the communication overhead grows exponentially.
2. **Criticality:** What happens if the system fails? Cockburn identified four levels:
    - **C (Comfort):** Loss of administrative comfort.
    - **D (Discretionary Money):** Loss of money that doesn't threaten the company.
    - **E (Essential Money):** Potential bankruptcy or massive financial loss.
    - **L (Life):** Potential loss of human life.

By mapping these dimensions, Crystal shows that the "darker" the color, the heavier and more disciplined the process needs to be to maintain safety, coordination, and reliability.

10.3. Choosing the Right Crystal Variant (Clear, Yellow, Orange, Red)

Crystal variants are named after colors, similar to gemstones, where the "clarity" or "density" changes based on the project's needs.

- **Crystal Clear (1–8 people):** The most popular variant. It focuses on "Osmotic Communication"—the idea that team members in the same room pick up information just by being near each other. It requires only three artifacts: a release plan, a list of items to be done, and status notes.
- **Crystal Yellow (10–20 people):** Introduces more formal roles, particularly around automated testing and clear ownership of components.
- **Crystal Orange (21–50 people):** Designed for medium-sized projects. It introduces "Reflection Workshops" every few weeks and requires more formal documentation for cross-team coordination.
- **Crystal Red (50–100+ people):** A heavy-weight methodology for large organizations where traditional Agile communication starts to break down.

10.4. Why It Faded—and What Persisted

Crystal is rarely implemented by name today because it is, by design, anti-standardization. In an era when companies seek a single model—like Spotify or SAFe—to scale across the organization, Crystal's insistence that each team defines its own process is often a hard sell for management.

However, Crystal's DNA is everywhere in modern Agile. The concept of **Osmotic Communication** is the foundation of co-located team rooms, and the **Reflection Workshop** was the direct ancestor of the **Sprint Retrospective**. Crystal reminds us that when a process feels too heavy, it likely is—and should be adapted accordingly.

10.5. Framework Operational Profile

To provide a concise reference and enable comparison with other frameworks, Crystal's practices are summarized in the Operational Profile below.

Table 10-1: Framework Operational Profile: Crystal

Feature	Description
1. Agile Approach Category	**Human-Centric / Contextual**
2. Core Intent / Purpose	To achieve project success through high-bandwidth communication and minimal process weight.
3. Primary Focus	People & Communication: Reducing the cost of information transfer between individuals.
4. Roles	Sponsor, Lead Designer, Programmer, User (Minimalist roles).
5. Cadence / Timing Model	**Continuous / Iterative:** Frequent delivery of working code (every 1–3 months).
6. Core Practices / Ceremonies	Osmotic Communication, Frequent Delivery, Reflective Improvement, Personal Safety.
7. Key Artifacts	Release Plan, User Stories/Requirements, Status Notes.
8. Work Management Model	**Empowered Teams:** Teams decide their own work-cycle and documentation levels.
9. Primary Metrics	**Delivery Frequency**, **Team Morale**, **Safety/Stability**.
10. Organizational Fit	Small, co-located teams or high-trust organizations.
11. Strengths	Extremely low overhead; high team autonomy; adaptable to any criticality level.
12. Common Failure Modes	**Lack of Structure:** Without a strong Lead Designer, the lack of prescriptive rituals can lead to a lack of focus.
13. Best Used When...	The team is small, highly skilled, and the priority is speed and developer happiness over formal reporting.

Crystal Family | Human-Centric

Identity

Criticality
Team Size

Crystal Family, 2004, Situational methodology family.

Core Intent

To match process weight to team size and system criticality.

Operational Heart

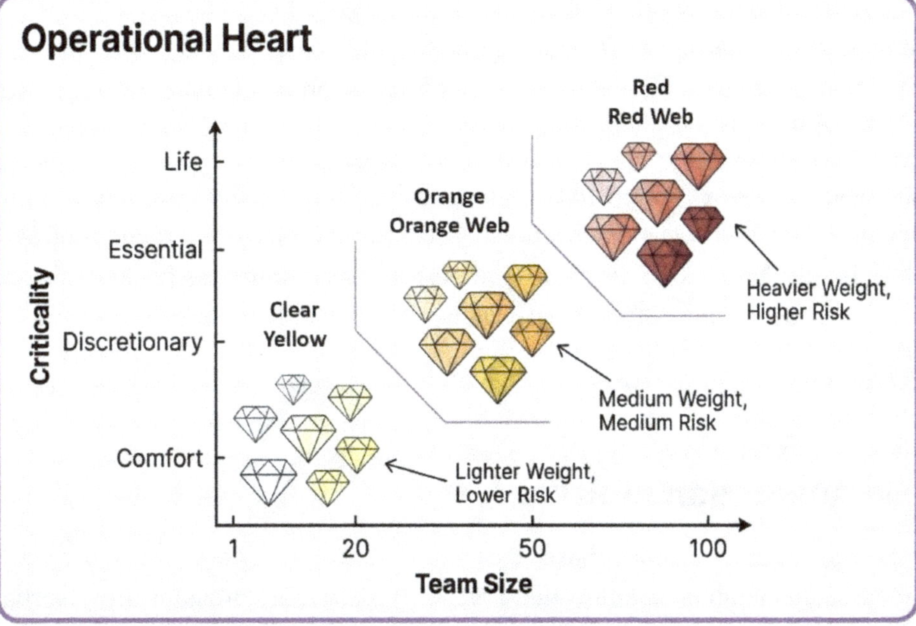

Process Lists

1. Release Planning
2. Reflection Workshops
3. Frequent Delivery
4. Osmotic Communication

Role Matrix

Sponsor (Funding)

Lead Designer (Technical guide)

Programmer (Builder)

Artifacts & Metrics

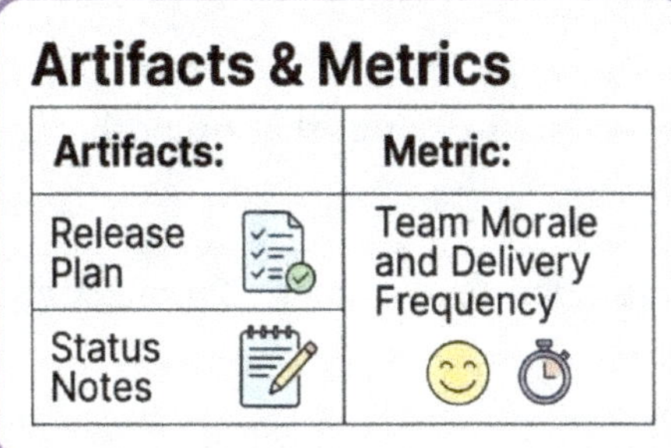

Artifacts:	Metric:
Release Plan	Team Morale and Delivery Frequency
Status Notes	

Context & Red Flags

Where it succeeds:

Co-located, high-trust skilled teams.

Red Flag:

Lack of focus without a strong Lead Designer.

Figure 10-1: Crystal Family (Human-Centric) Framework Snapshot

Chapter 11: Agile Unified Process (AUP) – The Bridge to Discipline

11.1. Origin and Ideology

The Agile Unified Process (AUP) holds a critical place in Agile history as the "middle ground" between the heavy, rigid processes of the 1990s and the lightweight agility of Scrum.

- **The Origin:** Developed by Scott Ambler (2005), AUP was a simplified version of IBM's Rational Unified Process (RUP). RUP was famously complex, often described as "Waterfall disguised as iterative." Ambler stripped away the excessive documentation and bureaucracy of RUP, applying Agile values to create a streamlined approach.
- **The Ideology:** AUP pioneered the philosophy of being **"Serial in the Large, Iterative in the Small."** It acknowledges that while developers work in fast cycles (Sprints), the organization still needs a clear beginning, middle, and end to its investment. Its ideology emphasizes full lifecycle coverage—ensuring that architecture, database management, and deployment receive as much discipline as coding itself.

11.2. Agile Adaptation of RUP

AUP organizes work into four distinct **Phases** that happen linearly (Serially), while the work within those phases happens through repeated **Iterations** (Iteratively). This structure provides the governance that predictive managers often miss in Scrum.

11.2.1. The Four Phases

1. **Inception:** Defining the project scope, identifying key risks, and gaining funding. (Goal: Agreement).
2. **Elaboration:** Defining the architecture and proving it works. (Goal: Risk Reduction).
3. **Construction:** The bulk of the development work, building the system iteratively. (Goal: Capability).
4. **Transition:** Testing, deployment, and training users. (Goal: Product Release).

11.2.2. The Disciplines

Unlike Scrum, which focuses on "events," AUP defines seven **Disciplines** that occur throughout these phases:

- Model, Implementation, Test, Deployment, Configuration Management, Project Management, and Environment.

11.3. Lifecycle Coverage and Enterprise Fit

AUP is rarely mentioned in "startup" circles because it feels too structured. However, its focus on **Enterprise Fit** makes it a vital case study for predictive managers:

- **Full Lifecycle View:** Unlike most frameworks, which begin with a backlog and end when code is delivered, AUP covers the project from idea inception through system retirement.
- **The Architect's Framework:** AUP emphasizes **Agile Modeling**, ensuring that teams understand the domain and technical constraints before heavy development begins.

11.4. Current Status: The Evolution into Disciplined Agile

It is important to note that **AUP is considered a deprecated framework** in the modern landscape.

- **Superseded by DA:** In 2012, **Scott Ambler and Mark Lines (2020)** evolved AUP into Disciplined Agile Delivery (DAD), expanding the focus beyond software to the entire business context. He recognized that AUP focused too heavily on software development and not enough on the broader business context (DevOps, IT operations, and enterprise architecture).
- **Why It Faded:** The rise of Continuous Delivery made the separate "Transition" phase (Phase 4) largely obsolete for digital products, as modern teams release code daily rather than once at the end of a project.
- **What Survived:** AUP's DNA is the foundation of **Disciplined Agile (Chapter 17)**. The concept of "Serial in the Large, Iterative in the Small" remains the dominant model for large-scale physical engineering (hybrid agile) and government projects where funding is phased.

11.5. Framework Operational Profile

To provide a concise reference and enable comparison with other frameworks, AUP's practices are summarized in the Operational Profile below.

Table 11-1: Framework Operational Profile: AUP

Feature	Description
1. Agile Approach Category	**Historical / Hybrid** (Precursor to Disciplined Agile)
2. Core Intent / Purpose	To simplify the heavy RUP framework for agile adoption.
3. Primary Focus	**Lifecycle Coverage:** Managing the project from concept to retirement.
4. Roles	Simplified RUP roles (Modelers, Implementers, Project Manager).
5. Cadence / Timing Model	**Phased & Iterative:** Four serial phases (Inception to Transition) containing iterative sprints
6. Core Practices / Ceremonies	Agile Modeling, Test-Driven Development (TDD), Iterative Development, Database Refactoring.
7. Key Artifacts	Domain Model, Architecture Document (Lightweight), Release Plan.
8. Work Management Model	**Phase-Gate Agile:** Progress is measured by passing milestones at the end of each phase (e.g., "Lifecycle Architecture Milestone").
9. Primary Metrics	Phase Milestones, Risk Mitigation, Defect Trends.
10. Organizational Fit	**Legacy:** Mostly found in organizations transitioning from RUP or utilizing Hybrid-Waterfall models.
11. Strengths	Bridges the gap for traditional project managers; strong focus on architecture.
12. Common Failure Modes	**Obsolescence:** Using AUP instead of modern Disciplined Agile (DA) or DevOps practices; treating the phases as strict Waterfall gates.
13. Best Used When...	You are studying the history of Agile scaling, or maintaining legacy systems that were built using RUP governance.

Agile Unified Process | AUP

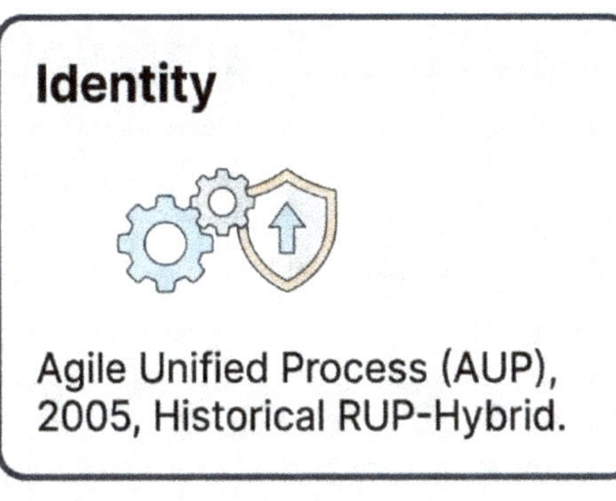

Core Intent

To bridge traditional discipline with Agile delivery loops.

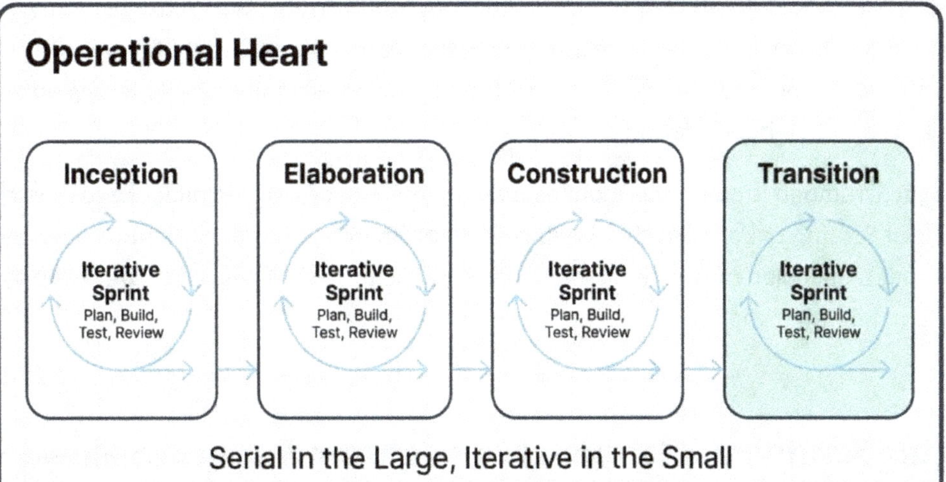

Process Lists

1. Inception
2. Elaboration
3. Construction
4. Transition

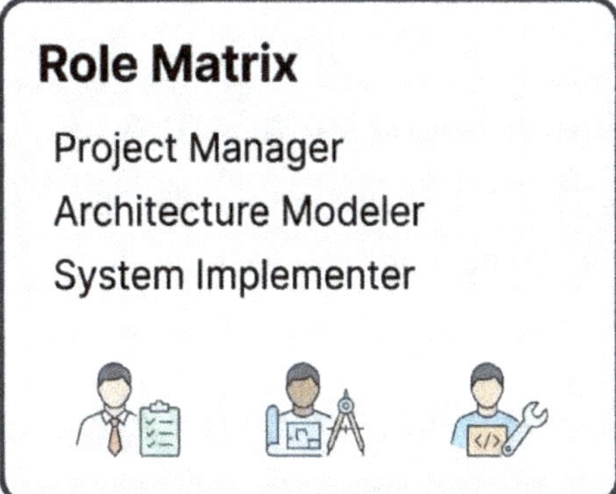

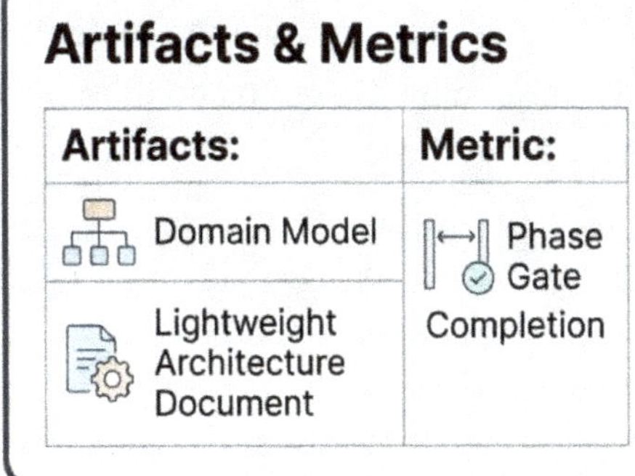

Figure 11-1: Agile Unified Process (AUP) Framework Snapshot

Chapter 12: Scrumban — The Hybrid Flow Framework

12.1. Origin and Ideology

Scrumban is the ultimate "evolutionary" framework, born from the practical reality that some work environments are too volatile for fixed-length Sprints.

- **The Origin:** Developed by **Corey Ladas** (2009) in his book, *Scrumban: Essays on Kanban Systems for Lean Software Development,* Originally meant to help teams transition from Scrum to Kanban, it quickly became a standalone framework for teams that found "pure" Scrum too restrictive and "pure" Kanban too loose. Corey Ladas recommends that teams adopt Scrumban gradually: start with existing Scrum roles and boards, then transition to flow-based planning as they gain experience with WIP limits and pull-based delivery.
- **The Ideology:** Scrumban is built on the philosophy of **"Pull over Push."** While it keeps the roles and some rituals of Scrum, it abandons the timeboxed iteration in favor of a continuous flow. Its core ideology is that teams plan only when they have the capacity. Work is pulled through the system based on demand rather than following a pre-defined calendar.

12.2. Bridging Scrum and Kanban: Moving from Sprints to Flow

Scrumban acts as a bridge, taking the best of both worlds to create a system that is both predictable and highly responsive.

12.2.1. What it keeps from Scrum

Teams often keep the **Daily Stand-up** and the **Retrospective**. More importantly, they often maintain the **Scrum Roles** (Product Owner and Scrum Master/Coach), though the focus shifts from "managing the Sprint" to "managing the flow."

12.2.2. What it adopts from Kanban

Scrumban replaces the Sprint Backlog with a **Kanban Board** featuring strict **WIP Limits**. The most significant change is the removal of Sprint Planning. In Scrumban, planning is **trigger-based**: when the "To Do" column drops below a predefined order point, the team pulls in the Product Owner to replenish the queue. Ladas emphasizes the use of **strict WIP limits** to prevent overloading the team and to make bottlenecks visible, which he identifies as a core principle for maintaining flow.

12.3. Continuous Delivery Without Prescribed Iterations

Scrumban excels in environments with **unpredictable work** without losing the strategic alignment of a product roadmap.

- **No Sprint Pressure:** Features are completed when truly done, reducing technical debt and stress.
- **Continuous Prioritization:** The Product Owner can dynamically adjust priorities without disrupting ongoing work. Ladas emphasizes that teams should **focus on completing features to the definition-of-done standard** rather than rushing to arbitrary deadlines.

Niche Success Scenarios:

These characteristics make Scrumban particularly effective in environments where work is unpredictable or priorities change frequently:

- **Maintenance and Support Teams:** Ideal for emergency work, where fixed Sprints are impractical.
- **Mature Product Teams:** Ladas suggests adopting Scrumban after mastering TDD and CI/CD, as it reduces planning overhead while maintaining flow.
- **Research & Development Teams:** Works well where task durations are highly uncertain.

12.4. Framework Operational Profile

To provide a concise reference and enable comparison with other frameworks, Scrumban's practices are summarized in the Operational Profile below.

Table 12-1: Framework Operational Profile: Scrumban

Feature	Description
1. Agile Approach Category	**Hybrid / Flow-Based**
2. Core Intent / Purpose	To provide a smooth transition from Scrum to Kanban or a permanent hybrid for unpredictable environments.
3. Primary Focus	**Flow & Capacity:** Minimizing planning overhead while maintaining team structure.
4. Roles	Typically retains Scrum roles (PO, Scrum Master/Coach, Team), but they are less prescriptive.
5. Cadence / Timing Model	**Continuous Flow:** No Sprints; releases happen when a feature is ready or on a fixed calendar date.
6. Core Practices / Ceremonies	Daily Stand-up, Retrospective, Trigger-based Replenishment (On-demand planning).
7. Key Artifacts	Kanban Board with WIP Limits, Cumulative Flow Diagram (CFD).
8. Work Management Model	**Pull System:** Work is pulled into the "Ready" column based on a re-order point.
9. Primary Metrics	**Lead Time**, **Cycle Time**, **Throughput**.
10. Organizational Fit	Maintenance teams, DevOps, and mature Agile teams looking for lower ceremony overhead.
11. Strengths	High flexibility; reduced planning waste; graceful handling of emergency work.
12. Common Failure Modes	Loss of focus without Sprint deadlines; skipping Retrospectives.
13. Best Used When…	The team is experienced and the nature of the work is too volatile for stable 2-week commitments.

Scrumban | Hybrid Flow

Identity

Scrumban (Hybrid Flow), 2009, Evolutionary Flow framework.

Core Intent

To bridge Scrum structure with Kanban responsiveness.

Operational Heart

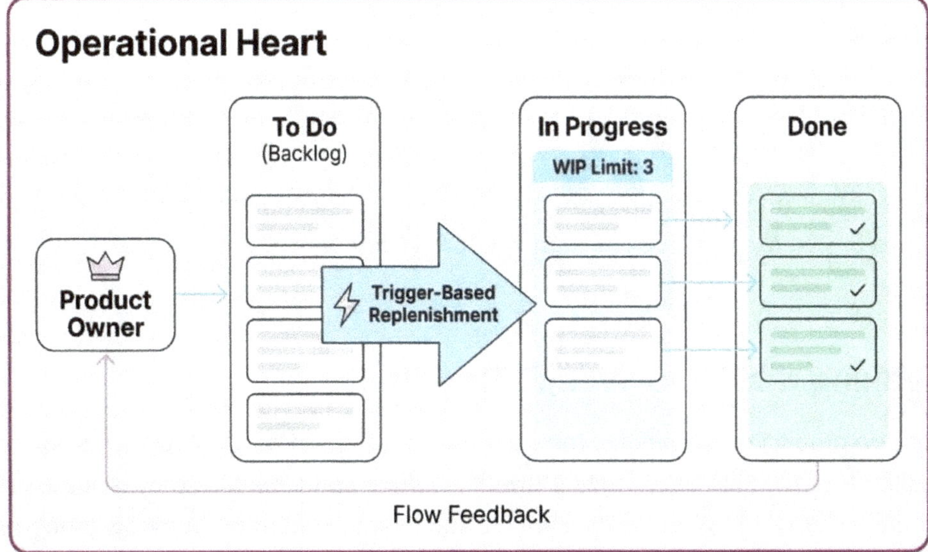

Process Lists

1. Daily Stand-up	
2. On-Demand Planning	
3. Service Review	
4. Flow Retrospective	

Role Matrix

Product Owner (Value)

Flow Coach (formerly SM, Effectiveness)

The Self-Managing Team (Delivery)

Artifacts & Metrics

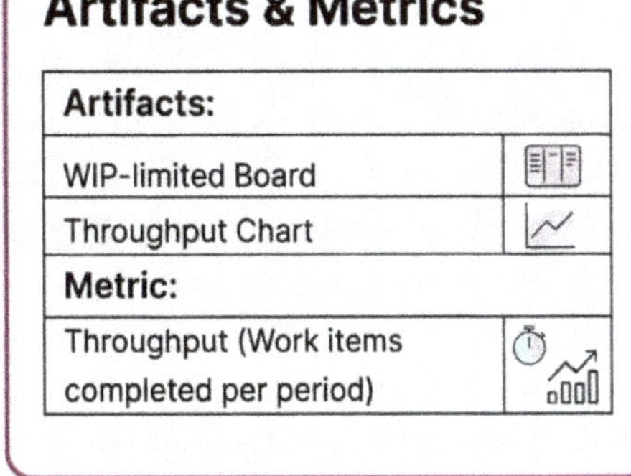

Artifacts:	
WIP-limited Board	
Throughput Chart	
Metric:	
Throughput (Work items completed per period)	

Context & Red Flags

Where it succeeds:
Mature teams or maintenance/support.

Red Flag:
Losing strategic alignment without iteration deadlines.

Figure 12-1: Scrumban (Hybrid Flow) Framework Snapshot

Chapter 13: Shape Up: The Product-Led Alternative

13.1. Origin and Ideology

- **The Origin:** Shape Up was formalized by **Ryan Singer** (2019), encapsulating over 15 years of product development practices at Basecamp. (formerly 37signals). It emerged as a reaction against the "Scrum treadmill"—the feeling of endless tickets and backlogs that never seem to shrink. Basecamp realized that standard Agile practices often led to "shipping nothing" because teams were stuck in a cycle of constant re-prioritization and micro-management.
- **The Ideology:** Shape Up creates a dual-track system: one track for "Shaping" (strategic definition) and one for "Building" (execution). Its core ideology is **Fixed Time, Variable Scope**. Instead of estimating how long a feature will take (which is usually wrong), Shape Up asks, "How much time is this idea worth?" (the Appetite). Teams are then given full autonomy to adjust the scope of the solution to fit that fixed timebox. It shifts the focus from "velocity" to "shippability."

13.2. Breaking the "Two-Week" Treadmill

For many mature product organizations, the standard Scrum rhythm—the "two-week sprint"—eventually feels less like agility and more like a treadmill. Developers often complain that they lack the time to think deeply about complex problems because they are constantly interrupted by "Backlog Refinement" meetings or the pressure to deliver small increments every ten days.

Shape Up, a methodology pioneered by Basecamp, offers a radical alternative. It rejects the idea of a never-ending backlog and the constant "slicing" of work into tiny user stories. Instead, it prioritizes **deep work** and **finished products** over constant status updates.

13.3. The 6+2 Rhythm: Cycles and Cool-downs

The heartbeat of Shape Up is not two weeks, but a structured **eight-week block**.

- **The Six-Week Build Cycle:** This duration is long enough to build something meaningful (a whole feature) from start to finish, but short enough to feel the deadline looming. This pressure compels the team to make active trade-offs—what Shape Up calls **"hammering the scope."**
- **The Two-Week Cool-down:** After every six-week cycle, there is a two-week "Cool-down" period. This is a buffer where there is no scheduled work. Teams use this time to fix bugs, refactor code, or explore new ideas freely.
- **The Betting Table (Concurrent):** While the builders are in "Cool-down," leadership meets at the **Betting Table** to decide what to build in the *next* cycle. This ensures that the "Shaping" track remains asynchronous and ahead of the "Building" track.

For the predictive manager, this offers more stability than Scrum. Instead of the scope changing every two weeks, the team is given six weeks of uninterrupted time to solve a specific problem within a **Fixed Appetite**.

13.4. Shaping: The Strategic Guardrail

In traditional Agile, "Backlog Refinement" is often a collaborative activity that can lead to "design by committee." In Shape Up, work is **Shaped** by a small group of senior leaders or designers before it is ever given to the team.

"Shaping" is the process of defining the solution at the right level of abstraction. It is:

1. **Rough:** It's not a high-fidelity mockup (which removes creativity). It is a "fat marker sketch" that outlines the boundaries.
2. **Solved:** The major technical risks and open questions are resolved by senior staff upfront. The team is not asked to "figure out if this is possible"; they are asked to "build this within the boundaries".
3. **Bounded:** The appetite (budget) is set. "We are willing to spend six weeks on this, but no more."

This resolves a common friction point: it allows senior leaders to set the strategy and scope (Shaping) while leaving the detailed implementation decisions entirely to the team.

13.5. The Betting Table: No More Backlogs

Perhaps the most controversial aspect of Shape Up is that **it eliminates the Backlog**.

In traditional Agile, backlogs grow indefinitely, becoming a "graveyard of good ideas" that everyone knows will never get built. Shape Up replaces this with the **Betting Table**.

- **The Pitch:** If someone wants a feature built, they must write a "Pitch" that defines the problem, the shaped solution, and the appetite.
- **The Bet:** Every six weeks, leadership sits at the Betting Table to review Pitches. They place a "Bet" on a few pitches to fill the next cycle.
- **The Wipe:** Critical to this method is that **pitches that are not chosen are discarded**. They do not go into a backlog. If the idea is truly important, it will be pitched again next time.

This approach aligns perfectly with Lean thinking by eliminating the "inventory" of stale tickets that clutter Jira and distract the team.

13.6. Monitoring Progress: The Hill Chart

Shape Up replaces the "Burn-down Chart" (which tracks tasks) with the **Hill Chart** (which tracks certainty).

- **The Uphill (Figuring it out):** The team is still discovering the "how." There is high uncertainty.
- **The Top of the Hill:** The "I've thought it through" moment. No more major unknowns.
- **The Downhill (Making it happen):** Execution. Pure production.

It allows stakeholders to see where a project is "stuck" in the problem-solving phase rather than just seeing a count of completed tickets.

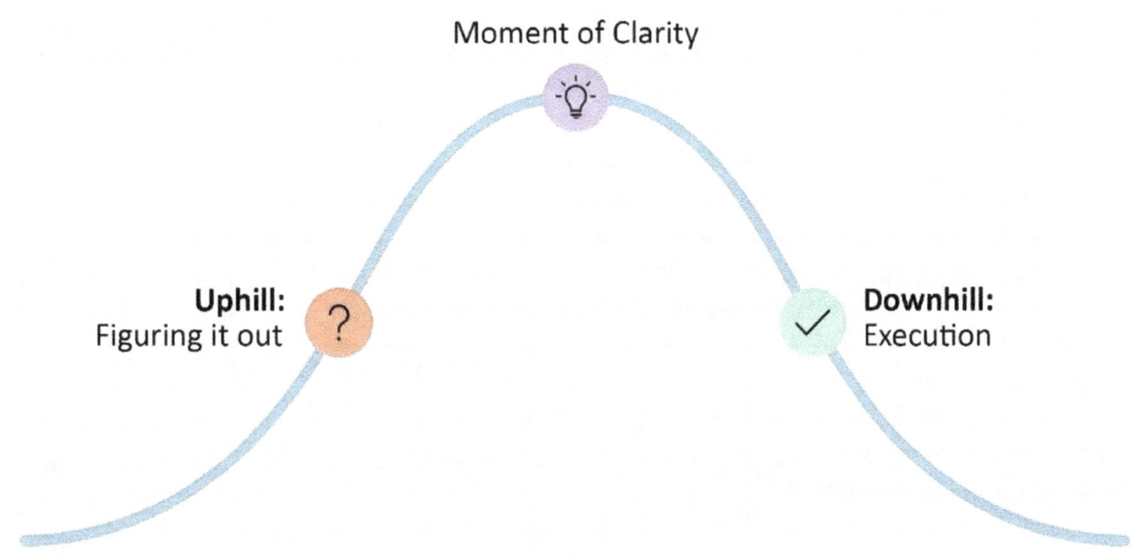

Figure 13-1: Shape Up Hill Chart: Tracking the Unknown

13.7. The Circuit Breaker: Managing Risk

The **Circuit Breaker** is the ultimate governance tool in modern Agile ecosystem. If a project is not finished by the end of the six-week cycle, it is **cancelled by default**. There are no "automatic extensions."

This forces the team to manage scope aggressively and prevents the "Sunk Cost Fallacy." If the work was truly valuable but unfinished, it must be re-pitched and re-bet upon at the next table.

13.8. Framework Operational Profile

To provide a concise reference and enable comparison with other frameworks, Shape Up's practices are summarized in the Operational Profile below.

Table 13-1: Framework Operational Profile: Shape Up

Feature	Description
1. Agile Approach Category	Product-Led / Value Stream Orchestration
2. Core Intent / Purpose	Fixes time and varies scope to break the constant sprint cycle and ensure meaningful shipping.
3. Primary Focus	**Dual-Track Delivery**: Asynchronous Shaping (strategic) and Building (execution).
4. Roles	**Shapers** (leaders defining work) and **Builders** (cross-functional teams executing work).
5. Cadence / Timing Model	**8-Week Block**: 6-Week Execution Cycle + 2-Week Cool-down (maintenance/breather).
6. Core Practices / Ceremonies	Shaping (pre-work), Betting Table (selection), Kick-off, Hill Charts (tracking). and The Circuit Breaker
7. Key Artifacts	The Pitch (proposal), Fat Marker Sketches (lo-fi designs), and Hill Charts (visualizing unknowns).
8. Work Management Model	**Betting:** No backlogs. Work is selected for the cycle or discarded; scope is fitted to the timebox.
9. Primary Metrics	Hill Chart Progress (Uphill: figuring out; Downhill: execution) and Cycle Completion Rate.
10. Organizational Fit	Mature, high-trust product orgs with self-directing teams requiring no daily Project Manager.
11. Strengths	Removes backlog overhead, guarantees focus time, and aligns budget ("appetite") with scope upfront.
12. Common Failure Modes	**Mini-Waterfall:** Over-specifying solutions. **Failure to Ship:** Hesitating to cut scope to meet the 6-week deadline.
13. Best Used When...	Two-week sprints are too short for complex creative work, or backlog grooming becomes burdensome.

Shape Up | Product-Led

Identity

Shape Up (Product-Led), 2019, Deep Work framework.

Core Intent

To ensure shippability by betting on meaningful projects in 6-week cycles.

Operational Heart

Track 1: Shaping (Discovery)

Shape → Bet

Track 2: Building (Delivery)

Build → Cool Down

Process Lists

1. Shaping
2. Pitching
3. Betting
4. Building
5. Cool-down

Role Matrix

Shapers (Define boundaries)

Builders (Cross-functional team)

No day-to-day PMs

Artifacts & Metrics

Artifacts:
The Pitch
Hill Chart (Tracking unknowns)
Metric:
Cycle Completion Rate

Context & Red Flags

Where it succeeds:
High-trust product organizations.

Red Flag:
Over-specifying solutions (Mini-Waterfall) in Shaping.

Figure 13-2: Shape Up (Product-Led) Framework Snapshot

Part IV: Scaling Agile for the Enterprise

Chapter 14: Scaling Fundamentals

14.1. Why Scaling Is Hard: The Communication Explosion

The most common mistake in the Agile landscape is assuming that if one team is productive, ten teams will be ten times as productive. This is rarely the case due to **Combinatorial Complexity** (often referred to in management as the *Communication Channel Explosion*).

As the number of nodes (people or teams) in a system increases, the number of potential communication paths grows quadratically, not linearly. This phenomenon, often linked to Brooks' Law (Brooks, 1975), is mathematically defined in the *Agile Practice Guide* (Project Management Institute [PMI], 2021) by the formula:

$$n(n-1)/2$$

Where n is the number of people.

- A team of **5** has **10** channels.
- A group of **50** has **1,225** channels.

This exponential growth creates a "Communication Tax." As teams grow, an increasing percentage of a person's time is spent synchronizing information rather than delivering value. Scaling Agile is, at its core, an exercise in managing this tax by limiting the number of active channels required to get work done.

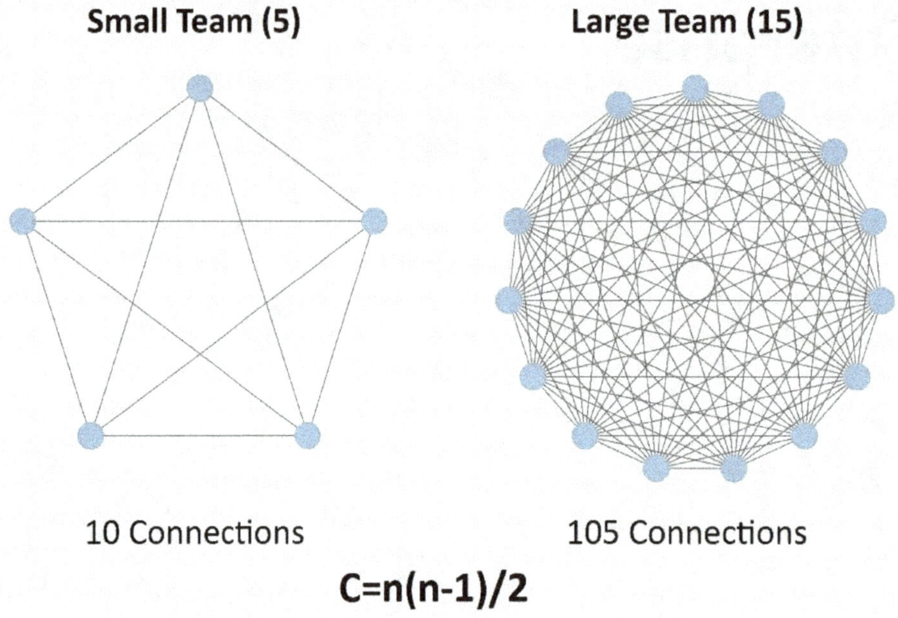

Figure 14-1: The Communication Tax in Scaling

14.2. Dependencies, Coordination, and Communication

When multiple teams work on the same product, they inevitably bump into one another. These "inter-team dependencies" are the "gravity" of scaling.

- **Technical Dependencies:** Team A cannot finish their feature until Team B provides an API.
- **Resource Dependencies:** Three teams all need the expertise of one specific Database Administrator (DBA).
- **Knowledge Dependencies:** One team holds the "tribal knowledge" required for others to proceed.

Effective scaling frameworks aim to solve these through **Coordination Mechanisms**. This might involve "Scrum of Scrums," "Big Room Planning," or "Communities of Practice." The goal is always to ensure that coordination costs never outweigh the value of the coordinated work.

14.3. Common Scaling Anti-Patterns

In *The Agile Landscape*, we often see organizations "failing to scale" because they apply the wrong mental models.

1. **Scaling a Broken Process:** If your team-level Scrum is dysfunctional, scaling it will only multiply the dysfunction.
2. **The "Copy-Paste" Model:** Taking a framework (like the "*Spotify Model*") and applying it to your organization without regard for your unique context.
3. **Command-and-Control Scaling:** Using a scaling framework as a new way to micromanage, insisting on centralized decision-making that slows the system down.
4. **Creating "Dependency Teams":** Organizing around specialized components (e.g., "Backend Team" and "UI Team") instead of end-to-end customer value, creating a nightmare of handoffs.

14.4. Scaling vs. Descaling: When Less Is More

The most provocative concept in modern Agile is that **the best way to scale is to "descale."** Descaling aligns with the principles of simplification and amplification required to wire a winning organization (Kim & Spear, 2023).

- **Scaling:** Adding more people, more layers of management, and more complex tooling to handle a massive project.
- **Descaling:** Reducing the complexity of the product, simplifying the architecture, and empowering small, cross-functional teams to own entire features from start to finish.

In the context-driven guide, we must ask: **"Do we really need more people, or do we need a better-defined product?"** Descaling reduces the "Communication Tax" by lowering the number of people/teams (n) needed to deliver a single unit of value.

The Scaling Paradox: The more you try to coordinate everyone with rules and roles, the slower they move. The more you decouple them so they *don't* have to coordinate, the faster you scale.

Chapter 15: Scrum of Scrums (SoS)

15.1. Origin and Ideology

Scrum of Scrums was not an afterthought; it was part of the original vision for making Scrum work at scale.

- **The Origin:** The concept was first implemented by **Jeff Sutherland and Ken Schwaber** (2020) in 1996 at Individual Inc. They realized that as they added more teams to their product development, the teams began to drift apart. They needed a "meta-Scrum" to ensure that the work of many teams resulted in a single, integrated product increment.
- **The Ideology:** The core philosophy of SoS is **fractal scaling**. It posits that the same patterns that make a single team successful—transparency, inspection, and adaptation—can be applied to a "team of teams." It treats a group of teams as if they were a single unit, creating a coordination layer focused on the space between teams rather than internal team work.

15.2. Multi-Team Coordination and the Ambassador Role

The heart of SoS is the **Scrum of Scrums Meeting**, a specialized synchronization event that mirrors the Daily Scrum but operates at a higher level of abstraction.

15.2.1. The Ambassador Role

Each team selects a representative—often called an **Ambassador**—to attend the SoS.

- **Who is it?** While many teams default to sending the Scrum Master, the "*Context-Driven*" approach suggests sending whoever is best suited to discuss the current dependencies (often a technical lead or a rotating team member).
- **The Responsibility:** The Ambassador is not there to "report status" to a manager; they are there to identify where their team's work overlaps or conflicts with others.

15.2.2. The Four Questions

In a standard Daily Scrum, we talk about individual tasks. In a Scrum of Scrums, the conversation shifts to team-level integration. The Ambassadors typically answer:

1. What has my team done since we last met that affects other teams?
2. What will my team do before we meet again that affects other teams?
3. What is slowing my team down that other teams could help with?
4. Are we about to put something in another team's way?

15.3. Strengths and Limitations of SoS

As with any tool in *The Agile Landscape*, SoS is highly effective in some contexts and dangerously insufficient in others.

15.3.1. Strengths

- **Low Overhead:** It requires no new software, no expensive certifications, and no massive organizational restructuring.
- **Real-Time Problem Solving:** It moves dependency management from a weekly report to a daily conversation.
- **Scalability:** It can be scaled further (a "Scrum of Scrum of Scrums") for very large organizations.

15.3.2. Limitations

- **The Status Update Trap:** Without strong facilitation, SoS quickly devolves into a boring status meeting where no real coordination happens.
- **Surface-Level Only:** SoS improves coordination but does not address deeper structural issues such as architectural misalignment, unclear product ownership, or conflicting strategic priorities.
- **"Telephone Game" Risk:** Information can get lost or distorted as it travels from the team to the Ambassador and then to the SoS meeting.

15.4. Framework Operational Profile

To provide a concise reference and enable comparison with other frameworks, Scrum of Scrums' practices are summarized in the Operational Profile below.

Table 15-1: Framework Operational Profile: Scrum of Scrums

Feature	Description
1. Agile Approach Category	**Lightweight Scaling / Coordination Pattern**
2. Core Intent / Purpose	To coordinate dependencies across multiple teams working on a single product.
3. Primary Focus	**Integration & Synchronization:** Ensuring the "whole" is greater than the sum of the parts.
4. Roles	**Ambassadors** (representatives from each team), **SoS Facilitator** (often a senior Scrum Master).
5. Cadence / Timing Model	**Daily or Frequent:** Typically happens immediately after the individual teams' Daily Scrums.
6. Core Practices / Ceremonies	The Scrum of Scrums Meeting, optional Scaled Backlog Refinement.
7. Key Artifacts	**Scaled Impediment Backlog**, Shared Release Plan.
8. Work Management Model	**Cooperative:** Teams remain autonomous but synchronize their "outputs" to ensure integration.
9. Primary Metrics	**Integration Frequency**, **Number of Cross-Team Dependencies**, **Time-to-Resolution for Cross-Team Blockers**.
10. Organizational Fit	mall to medium scaling (3–9 teams); organizations looking for minimal viable scaling before introducing heavier frameworks.
11. Strengths	Very easy to implement; preserves team autonomy; costs almost nothing to start.
12. Common Failure Modes	Becoming a "management reporting" meeting; lack of technical depth in discussions.
13. Best Used When...	Multiple teams are working on a single product and need a simple way to stay in sync.

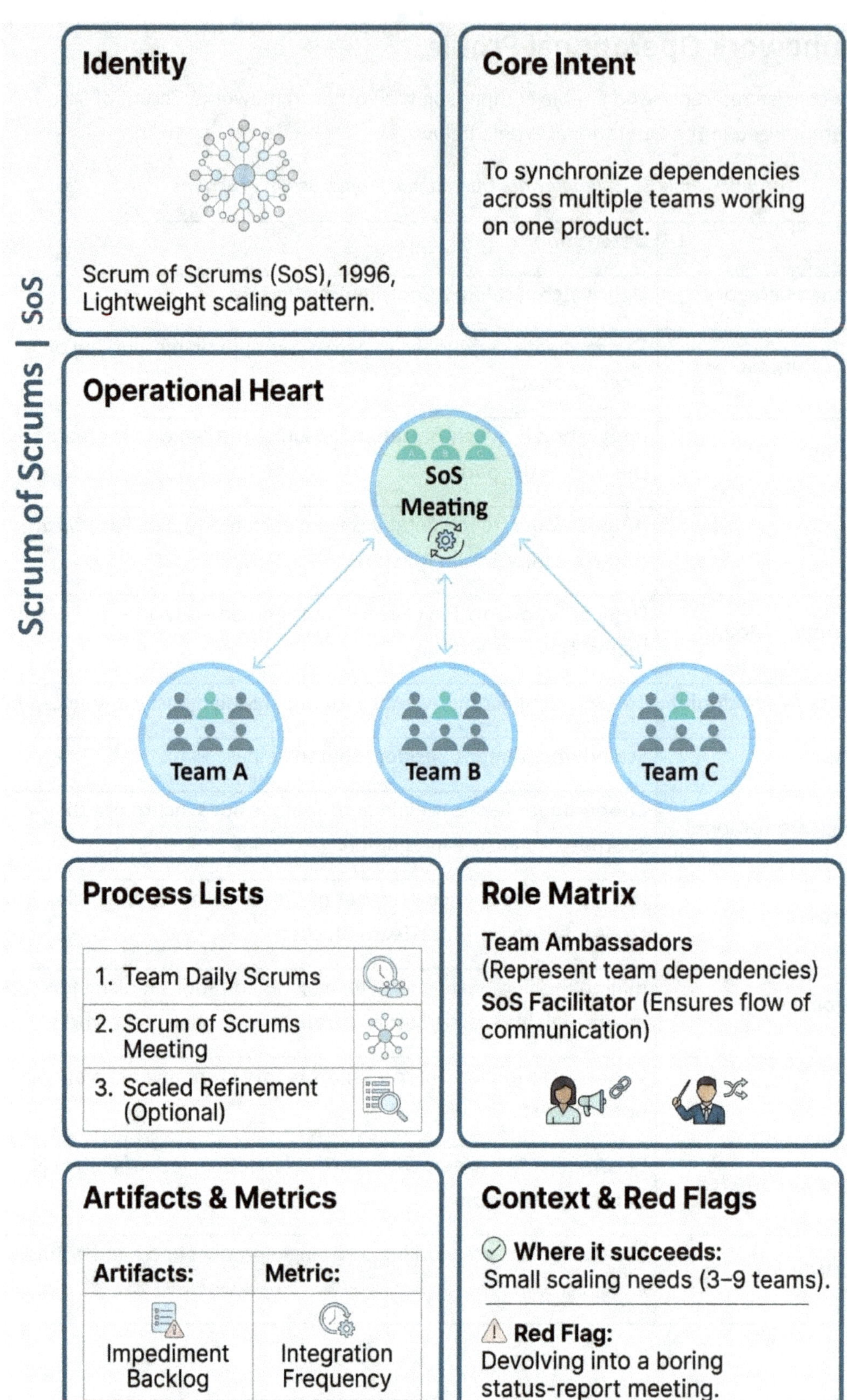

Figure 15-1: Scrum of Scrums (SoS) Framework Snapshot

Chapter 16: SAFe (Scaled Agile Framework)

16.1. Origin and Ideology

SAFe was designed to bridge the massive gap between executive-level strategy and team-level execution.

- **The Origin:** Created by **Dean Leffingwell** (Leffingwell et al., 2023) and launched in 2011, SAFe was built as a knowledge base of proven patterns for enterprise-scale development. It draws heavily from Lean, systems thinking, and Agile development, synthesizing them into a structured roadmap for large organizations.

- **The Ideology:** SAFe prioritizes organizational alignment in large enterprises, arguing that at scale, shared strategy, cadence, and intent are essential to avoid chaos. While team autonomy is encouraged at the local execution level, SAFe places stronger emphasis on alignment to the business mission and portfolio priorities. Its core ideology applies systems thinking to the organization itself—treating the enterprise as a complex system that must be optimized as a whole rather than as isolated teams.

16.2. Core Concepts: The Agile Release Train (ART) and PI Planning

To manage thousands of people, SAFe introduces several unique structural metaphors.

16.2.1. The Agile Release Train (ART)

The ART is the primary value-delivery mechanism in SAFe. It is a long-lived "team of teams" (typically 50–125 people) that are all synchronized on a common cadence.

- All teams on the "Train" start and end their iterations at the same time.

- The train is steered by roles like the **Release Train Engineer (RTE)**— often described as a “Chief Scrum Master”—and **Product Management**.

16.2.2. PI (Planning Interval) Planning (Program Increment Planning)

Formerly known as 'Program Increment' planning, the term was updated in **SAFe 6.0** (Leffingwell et al., 2023) to Planning Interval to better reflect the continuous nature of flow. The event remains the heartbeat of the Agile Release Train (ART), but the focus has shifted from "committing to a plan" to "aligning on a vision." Teams now utilize **Flow Accelerators**—such as visualizing WIP and minimizing handoffs—during planning to ensure the resulting plan is realistic rather than aspirational.

16.3. Benefits, Criticisms, and Trade-Offs

In *The Agile Landscape*, SAFe is often a polarizing topic. Understanding its context-driven trade-offs is essential for any practitioner.

16.3.1. Benefits

- **Strategic Alignment:** It is exceptionally good at ensuring that what a developer codes today is actually what the CEO promised the board six months ago.
- **Common Language:** It provides a standardized terminology that allows large, global organizations to communicate without confusion.
- **Dependency Management:** PI Planning is one of the most effective (albeit expensive) ways to solve complex cross-team dependencies.

16.3.2. Criticisms and Trade-Offs

- **The "Agile-in-Name-Only" Risk:** Critics argue that SAFe is simply "Waterfall with Sprints." The high level of upfront planning (PI Planning) can sometimes stifle the emergent nature of true Agility.
- **Complexity Overhead:** The framework is massive. Implementing it requires significant investment in training, certifications, and specialized roles.
- **Reduced Local Autonomy:** Teams often feel they have less freedom to pivot because they are "locked in" to a 10-week Program Increment plan.

16.4. Framework Operational Profile

To provide a concise reference and enable comparison with other frameworks, SAFe's practices are summarized in the Operational Profile below.

Table 16-1: Framework Operational Profile: SAFe

Feature	Description
1. Agile Approach Category	**Enterprise Scaling / Prescriptive Framework**
2. Core Intent / Purpose	To synchronize alignment, collaboration, and delivery for large numbers of Agile teams.
3. Primary Focus	**Program & Portfolio Alignment:** Connecting strategy to execution at scale.
4. Roles	RTE (Release Train Engineer), Product Management, System Architect, Business Owners.
5. Cadence / Timing Model	**Synchronized Iterations:** 2-week team sprints within a 10-week Program Increment (PI).
6. Core Practices / Ceremonies	PI Planning, System Demo, Inspect & Adapt (I&A), Scrum of Scrums.
7. Key Artifacts	Program Board, ART Backlog, PI Objectives, Architectural Runway.
8. Work Management Model	**Synchronized Pull with Planned Commitments:** Teams pull work from a shared Program Backlog during PI Planning.
9. Primary Metrics	**Program Predictability Measure**, **Feature Cycle Time**, **Flow Efficiency, Business Value Achievement**.
10. Organizational Fit	Large enterprises (500+ employees) with complex, inter-dependent products.
11. Strengths	Highly predictable; clear career paths; excellent for managing massive dependencies.
12. Common Failure Modes	**"The Red Tape Trap":** Adding too much bureaucracy; treating PI Plans as rigid contracts.
13. Best Used When...	You have hundreds of developers working on a single, highly complex system that requires strict alignment.

SAFe | Scaled Agile Framework

Identity

SAFe (Scaled Agile Framework), 2011, Enterprise Scaling/Prescriptive.

Core Intent

To align strategy and synchronization across hundreds of Agile teams.

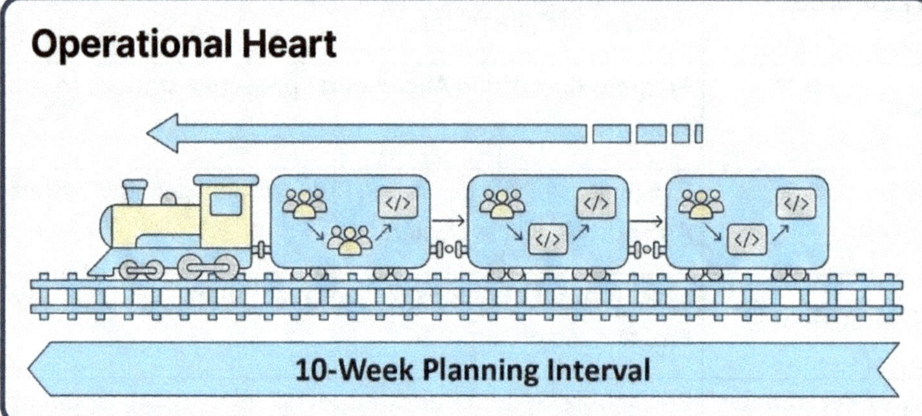

Process Lists

1. PI Planning	
2. System Demo	
3. Inspect & Adapt (I&A)	
4. Portfolio Sync	

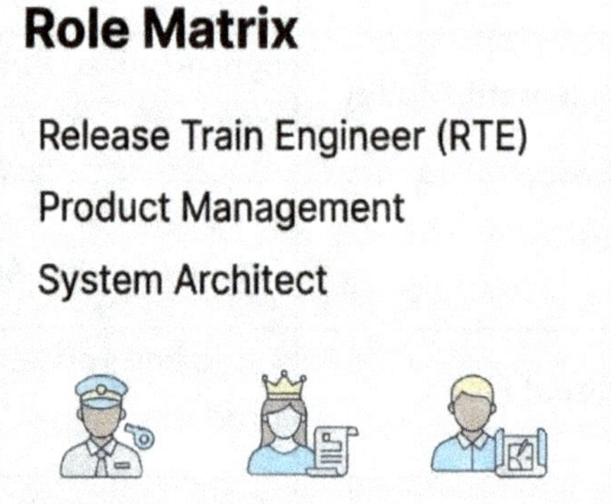

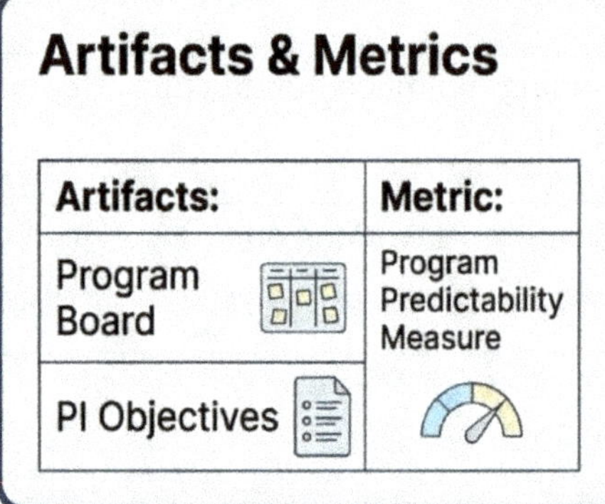

Artifacts:	Metric:
Program Board	Program Predictability Measure
PI Objectives	

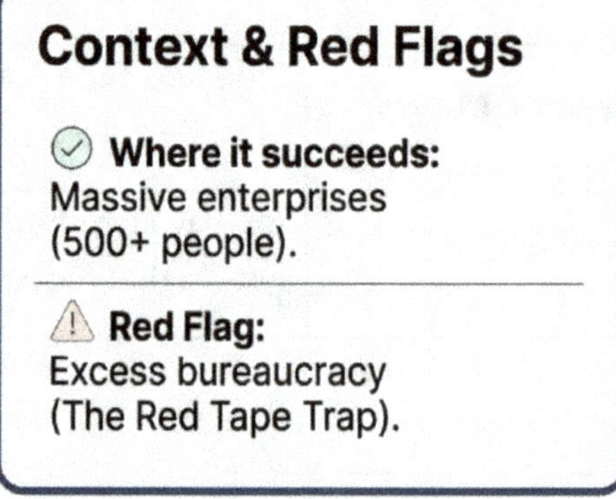

Figure 16-1: SAFe (Scaled Agile Framework) Framework Snapshot

Chapter 17: LeSS (Large-Scale Scrum)

17.1. Origin and Ideology

LeSS was born from a desire to keep Scrum "pure" even when the team count grows.

- **The Origin:** Developed by **Bas Vodde and Craig Larman** (2016) based on their experiences in large-scale telecoms and finance (specifically Nokia and Xerox) in the mid-2000s. They realized that most scaling problems were actually organizational design problems in disguise.
- **The Ideology:** LeSS is based on the principle that **"Scaling Scrum is Scrum."** It isn't a "new" framework; it is Scrum applied to multiple teams working on a single product. Its core ideology is **Empiricism** and **Lean Thinking**— relentlessly eliminating waste, especially the waste created by coordination-heavy roles and unnecessary management layers.

17.2. Minimalistic Scaling: Descaling the Organization

LeSS is famous for its "More with LeSS" mantra. It posits that to scale effectively, you should have:

- **More** learning, **Less** following of scripts.
- **More** value, **Less** waste.
- **More** whole-product focus, **Less** narrow-component focus.

17.2.1. Descaling vs. Scaling

Unlike prescriptive frameworks that add new roles (like Release Train Engineers), LeSS suggests that you should **descale** the organization. This means removing functional silos (like a separate QA or Architecture department) and instead creating "Feature Teams" that are capable of doing everything required to deliver a customer-centric feature from start to finish.

In LeSS, organizational change is not optional—it is the mechanism by which scaling is achieved.

17.3. Organizational Design and Product Focus

In the LeSS landscape, the structure of the company *is* the framework. It focuses on two major shifts:

17.3.1. One Product, One Product Owner, One Backlog

In many organizations, scaling leads to "Product Owner inflation," where every team has its own PO. LeSS forbids this. To maintain a true "Whole Product" view, there is only **one** Product Owner for up to eight teams. This forces the teams to collaborate directly with customers and stakeholders rather than relying on proxy roles to mediate customer contact.

17.3.2. Requirement Areas (LeSS Huge)

When a project grows beyond eight teams, LeSS evolves into **LeSS Huge**. Even then, it avoids adding hierarchy. Instead, it divides the product into **Requirement Areas** (customer-centric categories), each with its own Area Product Owner, but all still synchronized toward a single integrated product increment.

17.4. Framework Operational Profile

To provide a concise reference and enable comparison with other frameworks, LeSS' practices are summarized in the Operational Profile below.

Table 17-1: Framework Operational Profile: LeSS

Feature	Description
1. Agile Category	**Descaling / Minimalistic Scaling**
2. Core Intent	To apply the principles of Scrum to multiple teams without adding organizational complexity.
3. Primary Focus	**Whole Product Focus:** Ensuring all teams work on the most valuable items for the customer.
4. Roles	Product Owner (One per product), Scrum Master (typically 1–3 teams), Feature Teams.
5. Cadence	**Common Synchronous Sprint:** All teams start and end their Sprints at the same time.
6. Core Practices / Ceremonies	Overall Product Backlog Refinement, Sprint Planning One (Shared) & Two (Team), Overall Retrospective.
7. Key Artifacts	**One** Product Backlog, One Integrated Product Increment.
8. Work Management Model	**Self-Organizing Coordination:** Teams coordinate directly with each other (e.g., via "Scouts" or "Open Space").
9. Primary Metrics	**Lead Time, Cycle Time**, **Customer Value**, **Organizational Agility (ability to pivot)**.
10. Org Fit	High-trust environments; organizations willing to undergo radical structural change.
11. Strengths	Extremely low overhead; high transparency; eliminates silos; creates deep technical mastery.
12. Failure Modes	**Culture Shock:** Organizations that aren't ready to remove managers often find LeSS too "chaotic" or operationally "risky"
13. Best Used When...	You want the highest level of agility and are willing to change your organizational structure to get it.

LeSS | Large-Scale Scrum

Identity

Product Backlog

LeSS (Large-Scale Scrum), 2016, Descaling/Minimalistic scaling.

Core Intent

To apply pure Scrum to multiple teams by removing organizational bulk.

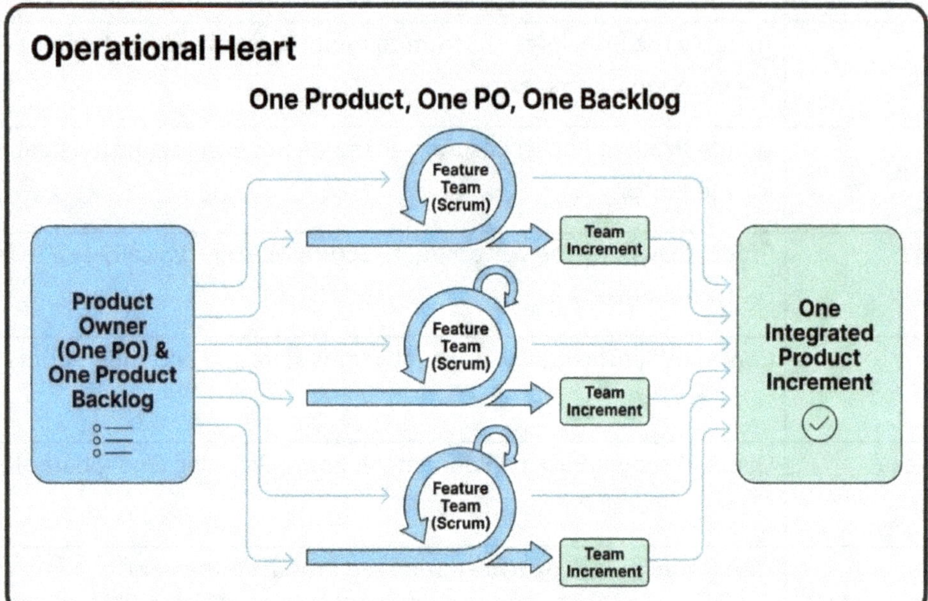

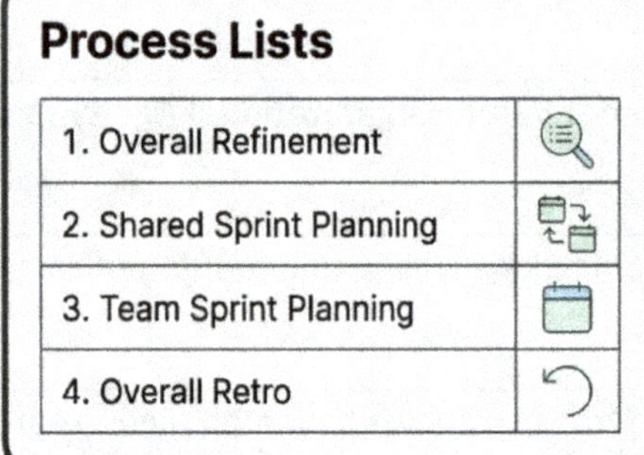

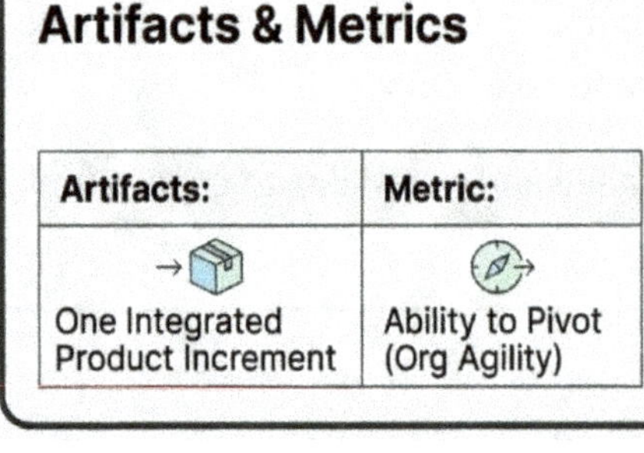

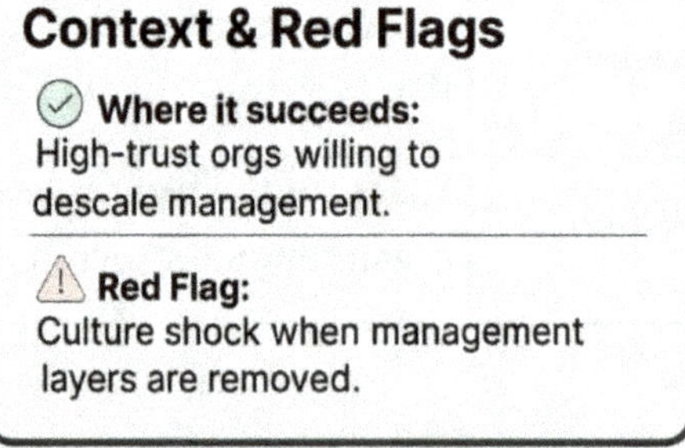

Figure 17-1: LeSS (Large-Scale Scrum) Framework Snapshot

Chapter 18: Disciplined Agile (DA)

18.1. Origin and Ideology

Disciplined Agile (DA) emerged in response to a recurring problem in large organizations: **“out-of-the-box” Agile frameworks often fail when applied rigidly to complex, real-world environments**.

- **The Origin:** Developed by **Scott Ambler and Mark Lines** (2020) and building on Ambler’s work with the Agile Unified Process, DA was eventually acquired by the Project Management Institute (PMI) in 2019. It was built as a "process decision toolkit" rather than a rigid framework. PMI’s acquisition matters for adoption because of increased credibility, training, and resources.
- **The Ideology:** The core philosophy of DA is **"Context Counts."** It assumes that every team is unique and faces a unique situation. Its goal is to provide a "professional's toolkit" that allows teams to make better-informed process decisions. It moves away from "Scrum-only" thinking to a broader, hybrid view that includes Lean, Kanban, and traditional techniques.

18.2. The Toolkit Mindset: Choose Your WoW

DA is famous for its focus on **'Choosing Your WoW' (Way of Working)**, a concept formalized in the official practitioner's handbook (Project Management Institute [PMI], 2022). Instead of telling you to "do a Daily Stand-up," it presents "Coordination" as a goal and offers several ways to achieve it.

18.2.1. Process Goals and Decision Blades

DA is organized into **Process Goals** (e.g., "Explore Scope," "Coordinate Activities," "Deploy the Solution"). For each goal, DA provides a **Decision Blade**—a list of techniques and practices with their associated trade-offs.

- **The Benefit:** It helps teams understand *why* they are choosing a practice, not just that they *must* do it.
- **The "Guided Evolution":** DA encourages teams to continuously improve by experimenting with different techniques from the toolkit, a process it calls **Continuous Guided Improvement (CGI)**.

18.3. The DA Lifecycles (Agile, Lean, Continuous Delivery, Exploratory)

One of the most powerful features of DA is that it doesn't force one "rhythm" on the organization. It recognizes that a marketing team, a software team, and a research team need different lifecycles. Teams select a lifecycle based on project complexity, uncertainty, and desired delivery cadence

DA offers six primary lifecycles to choose from:

1. **Agile Lifecycle:** Based on Scrum; best for new projects with high uncertainty.
2. **Lean Lifecycle:** Based on Kanban; best for high-volume, flow-based work.
3. **Continuous Delivery (Agile):** A more mature version of the Scrum-based lifecycle with frequent releases.
4. **Continuous Delivery (Lean):** A flow-based approach for teams that release code multiple times a day.
5. **Exploratory Lifecycle:** Based on Lean Startup method (Ries, 2011); best for R&D or "discovery" work where the goal is to validate a business idea.
6. **Program Lifecycle:** Designed for "teams of teams" to coordinate large-scale efforts.

18.4. Practical Guidance and Cautions

Disciplined Agile (DA) offers a flexible, context-driven toolkit, but its effectiveness depends heavily on how it is applied. The following points highlight practical guidance and common cautions:

18.4.1. Practical Guidance

- **Tailor with Care:** DA's strength lies in choosing the most appropriate practices for each team or situation. Take time to understand your context before selecting techniques from the toolkit.
- **Invest in Team Competence:** Teams must have sufficient Agile knowledge and decision-making capability to evaluate trade-offs effectively. DA is not a "plug-and-play" solution.
- **Encourage Continuous Improvement:** Use Continuous Guided Improvement (CGI) to experiment with different WoWs, learn from outcomes, and adjust practices over time.
- **Leadership Alignment:** Leaders should support experimentation, allow multiple valid approaches, and avoid enforcing a single "one-size-fits-all" methodology.

18.4.2. Cautions

- **Analysis Paralysis:** With many options available, inexperienced teams can become overwhelmed or indecisive. Start small and expand choices gradually. Teams should prioritize one or two process goals initially and expand as confidence grows.
- **Not a Fix for Dysfunction:** DA exposes organizational weaknesses but does not automatically resolve them. Structural or cultural issues must be addressed separately.
- **Consistency vs. Autonomy:** Balancing team autonomy with organizational alignment requires discipline; too much freedom without oversight can create confusion or misalignment—for example, teams may duplicate work.
- **Avoid Over-Tooling:** Implementing all aspects of the toolkit at once can create unnecessary complexity. Focus on practices that directly support current goals.

DA excels when teams and leaders embrace flexibility and context-driven decision-making, but it demands experience, discipline, and supportive organizational culture to realize its benefits. In short, DA works best in organizations willing to experiment, reflect, and adapt continuously.

18.5. Framework Operational Profile

To provide a concise reference and enable comparison with other frameworks, DA's practices are summarized in the Operational Profile below.

Table 18-1: Framework Operational Profile: DA

Feature	Description
1. Agile Category	**Hybrid Toolkit / Decision Framework**
2. Core Intent	To provide a comprehensive toolkit for choosing and evolving a tailored Way of Working (WoW).
3. Primary Focus	**Context-Driven Choice:** Giving teams the power to pick the best practices for their situation.
4. Roles	Team Lead (Agile Lead), Product Owner, Architecture Owner, Team Member, Stakeholder.
5. Cadence	**Variable:** Depends on the chosen lifecycle (Iterative, Flow, or Continuous).
6. Core Practices / Ceremonies	"Choosing Your WoW," Goal-driven retrospectives, Continuous Guided Improvement (CGI).
7. Key Artifacts	DA Toolkit/Process Goal Diagrams, Team WoW Handbook.
8. Work Management Model	**Contextual:** Can be backlog-driven, queue-driven, or experiment-driven.
9. Primary Metrics	**G6 (Group of 6) Metrics**, including Cycle Time, Throughput, and Value Delivered.
10. Org Fit	Organizations that value flexibility and want to move beyond "methodology dogma."
11. Strengths	Extremely flexible; vendor-neutral; acknowledges the reality of enterprise complexity.
12. Failure Modes	**Analysis Paralysis:** The sheer number of choices in the toolkit can overwhelm inexperienced teams.
13. Best Used When...	You have diverse teams with different needs and want a common language to help them all improve.

Identity

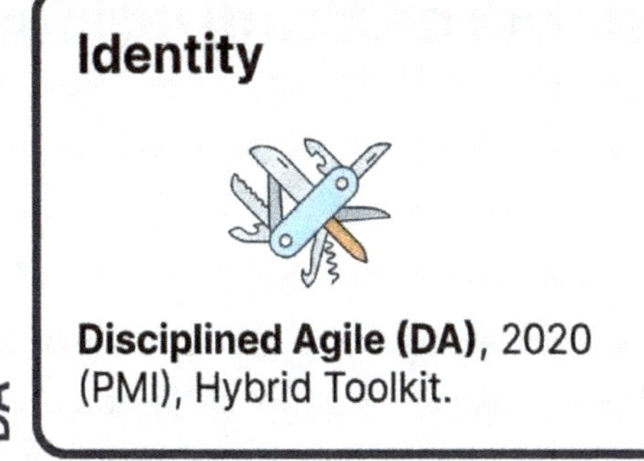

Disciplined Agile (DA), 2020 (PMI), Hybrid Toolkit.

Core Intent

To provide a comprehensive toolkit for choosing a tailored Way of Working (WoW).

Operational Heart

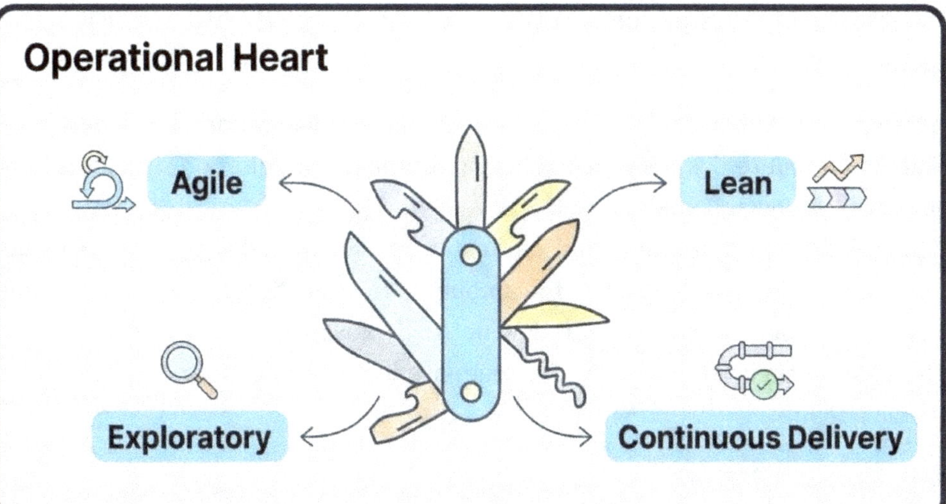

Process Lists

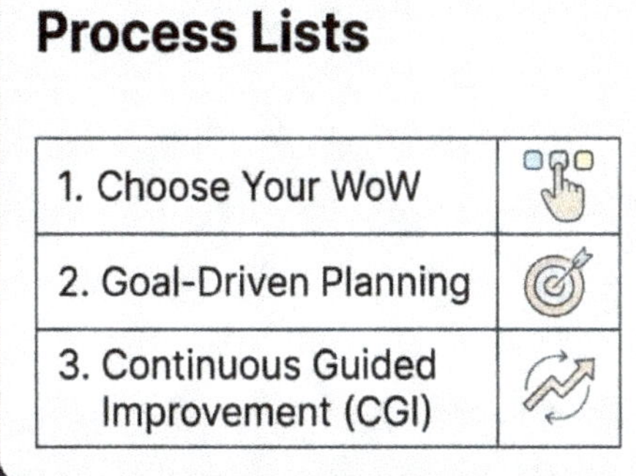

1. Choose Your WoW	
2. Goal-Driven Planning	
3. Continuous Guided Improvement (CGI)	

Role Matrix

Team Lead (Agile Lead)
Architecture Owner
Team Member

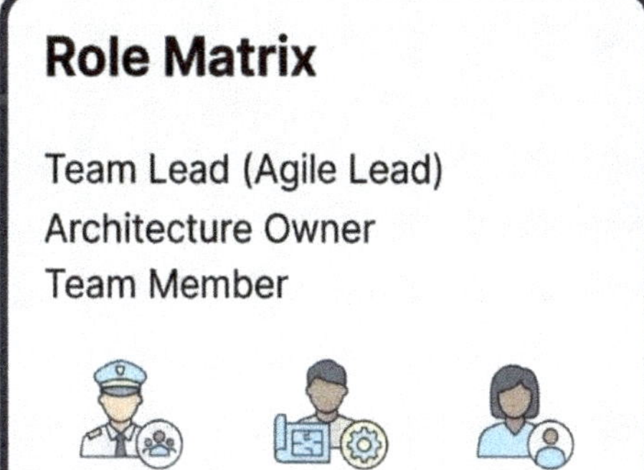

Artifacts & Metrics

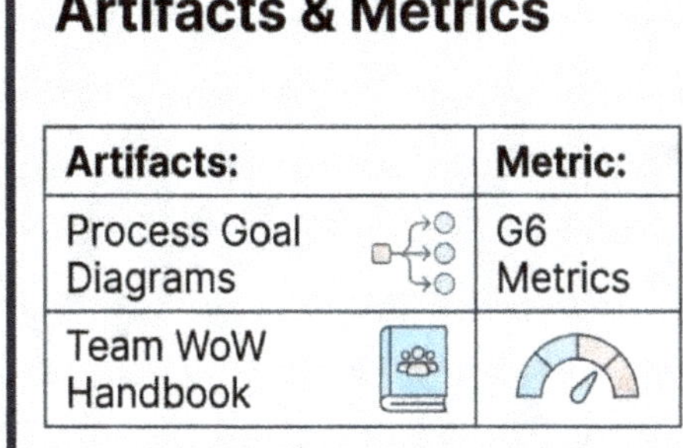

Artifacts:	Metric:
Process Goal Diagrams	G6 Metrics
Team WoW Handbook	

Context & Red Flags

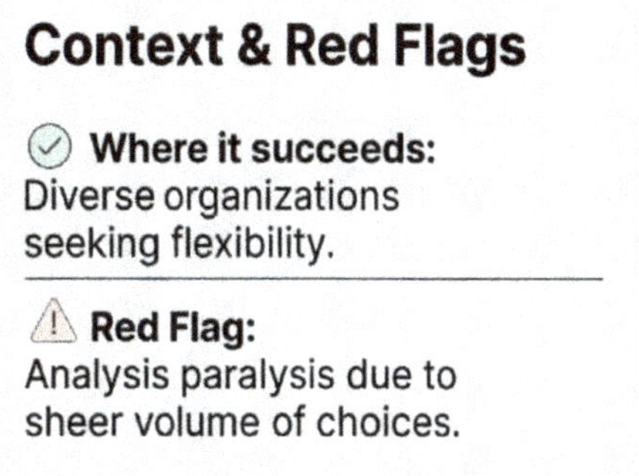

Where it succeeds: Diverse organizations seeking flexibility.

Red Flag: Analysis paralysis due to sheer volume of choices.

Figure 18-1Disciplined Agile (DA) Framework Snapshot

Chapter 19: Team Topologies: Modern Organizational Design

19.1. Origin and Ideology

- **The Origin:** Originally published in 2019 and updated for the era of AI-augmented work, **Matthew Skelton and Manuel Pais** (2025) developed Team Topologies as a response to the DevOps movement. The authors observed that many organizations were adopting tools (Kubernetes, Cloud, CI/CD) but failing to see results because their team structures were fighting against their software architecture.

- **The Ideology:** The framework is built on Conway's Law: "Organizations which design systems... are constrained to produce designs which are copies of the communication structures of these organizations." (Conway, 1968)
  Its ideology is that team structure must be treated as a software design concern. By optimizing for Fast Flow and limiting Cognitive Load (the amount of mental effort required by a team), organizations can build software that is decoupled, deployable, and manageable. It rejects the "Matrix" structure in favor of long-lived, stream-aligned teams supported by specialized platform and enabling teams.

19.2. Beyond Scaling: The Cognitive Load Constraint

For decades, the "scaling" conversation in Agile focused on coordination: *How do we get 50 teams to attend a meeting? How do we synchronize their backlogs?* Frameworks like SAFe and LeSS answer these questions with rigorous process overlays.

However, modern Agile recognizes that the fundamental constraint on speed is often not the process, but the **organizational structure** itself.

In a traditional matrix organization, teams are often overloaded. They are expected to know the business domain, the legacy codebase, the new cloud infrastructure, the security protocols, and the deployment pipeline. This state is known as **Cognitive Overload**.

When a team's cognitive load is exceeded, "flow" stops. Quality drops, motivation wanes, and bottlenecks emerge because the team is mentally saturated. *Team Topologies*, a modern paradigm popularized by Matthew Skelton and Manuel Pais, addresses this by shifting the goal of scaling from "adding more people" to **"optimizing cognitive load"**.

In the **Second Edition of *Team Topologies* (2025)**, Skelton and Pais elevate **Cognitive Load** from a constraint to a primary design principle. In an AI-augmented workforce, the risk is no longer just "too much work"—it is "context switching fatigue" caused by managing multiple AI tools and complex integrations.

Modern organizational design asks: "Does this team structure minimize the cognitive load required to deliver value?" If a Stream-aligned team is overwhelmed by infrastructure tasks, the organization is "architecturally broken," regardless of how hard the team works. The solution is not to hire more people, but to deploy a **Platform Team** to absorb that complexity via self-service APIs.

19.3. Conway's Law and the "Reverse Conway Maneuver"

To design an Agile organization, one must first understand **Conway's Law**, which states:

"Organizations which design systems are constrained to produce designs which are copies of the communication structures of these organizations."

In simple terms: if you have a siloed organization (separate Database, UI, and Logic teams), you will inevitably produce siloed, monolithic software that requires heavy coordination to change.

Modern organizational design applies the **"Reverse Conway Maneuver."** instead of letting the org chart dictate the architecture, we design the team structure to match the software architecture we *want*. If we want a modular, decoupled system that allows for fast releases, we must design modular, decoupled teams.

19.4. The Four Fundamental Team Topologies

To replace the chaos of "matrix management" with clear, flow-based interactions, *Team Topologies* proposes four specific team types. These are not job titles, but team behaviors.

19.4.1. Stream-aligned Teams

The **Stream-aligned Team** is the "heart" of the modern Agile organization. It is the primary value-delivery unit (similar to a Scrum team), but with a stricter definition: it is aligned to a single, continuous stream of work—such as a specific product, a user journey, or a service.

- **Goal:** Steady flow of value.
- **Characteristic:** These teams are cross-functional and empowered to deliver independently, without hand-offs to other departments.

19.4.2. Enabling Teams

In a complex environment, Stream-aligned teams cannot know everything. **Enabling Teams** are composed of experts (e.g., Technical Architects, Agile Coaches, or Security Specialists) whose sole purpose is to *upskill* Stream-aligned teams.

- **Goal:** Bridge capability gaps.
- **Anti-Pattern:** Enabling teams should not become "ivory towers" that dictate rules. Instead, they act as "servant-consultants," helping Stream-aligned teams acquire new skills so they can eventually be autonomous.

19.4.3. Complicated-subsystem Teams

Some parts of a system require such deep, specialized knowledge that asking a generalist Stream-aligned team to handle them would cause cognitive overload. A **Complicated-subsystem Team** is responsible for these highly specific components—such as a proprietary 3D rendering engine, a complex mathematical algorithm, or a face-recognition module.

- **Goal:** Encapsulate complexity.
- **Interaction:** They package this complexity into a module that Stream-aligned teams can simply "consume" without needing to understand the math inside.

19.4.4. Platform Teams

The **Platform Team** is the foundation that allows Stream-aligned teams to move fast. They treat the internal infrastructure as a product, providing a "Thinnest Viable Platform" (TVP) of self-service APIs, tools, and documentation.

- **Goal:** Reduce cognitive load for Stream-aligned teams.

- **The Shift:** Unlike traditional "Ops" teams that act as gatekeepers (requiring a ticket to provision a server), Platform teams build self-service tools so Stream-aligned teams can provision their own servers without waiting for permission. This is the essence of the "Internal Developer Platform" (IDP).

19.5. Interaction Modes

Identifying the teams is only half the battle; defining how they talk to each other is crucial to preventing the "Communication Explosion" discussed in Chapter 13.

- **Collaboration:** Working closely together for a defined period to discover new things (high interaction).
- **X-as-a-Service:** One team provides a service (like an API) that another consumes with minimal communication (low interaction).
- **Facilitating:** One team (usually Enabling) helps another team clear obstacles.

For the predictive project manager, *Team Topologies* offers a structural blueprint. It moves away from the vague notion of "cross-functional teams" toward a deliberate design where every interaction is planned to maximize flow and minimize the friction of hand-offs.

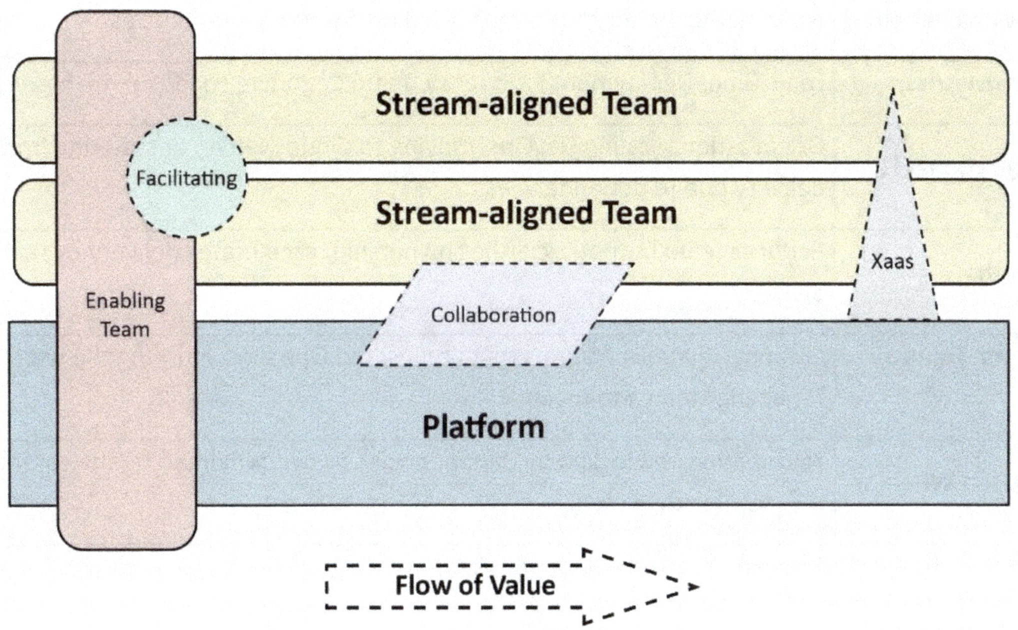

Figure 19-1: Team Topologies: Interaction Patterns

19.6. Framework Operational Profile

To provide a concise reference and enable comparison with other frameworks, Team Topologies' practices are summarized in the Operational Profile below.

Table 19-1: Framework Operational Profile: Team Topologies

Feature	Description
1. Agile Approach Category	Organizational Design / Structural Pattern
2. Core Intent / Purpose	To optimize the organization for fast flow of value by aligning team structures with software architecture.
3. Primary Focus	**Cognitive Load Management** & **Fast Flow**.
4. Roles (Team Types)	Stream-aligned, Enabling, Platform, Complicated-subsystem.
5. Cadence / Timing Model	Evolutionary; structure changes are triggered by flow bottlenecks, not fixed dates.
6. Core Practices	Interaction Modes (Collaboration, X-as-a-Service, Facilitating), Team API definition.
7. Key Artifacts	Team API, Thin Platform, Cognitive Load Assessments.
8. Work Management	Flow-based; aimed at minimizing hand-offs and dependencies.
9. Primary Metrics	Lead Time, Deployment Frequency, Team Cognitive Load (survey-based).
10. Organizational Fit	Organizations scaling DevOps, moving to Cloud-Native, or suffering from slow delivery due to dependencies.
11. Strengths	Reduces team burnout; clarifies ownership; accelerates delivery by removing blockers.
12. Common Failure Modes	Creating "Platform" teams that are just old Ops silos; ignoring the need for "Enabling" teams to teach skills.
13. Best Used When...	Your teams are blocked by dependencies or overwhelmed by the complexity of the systems they own.

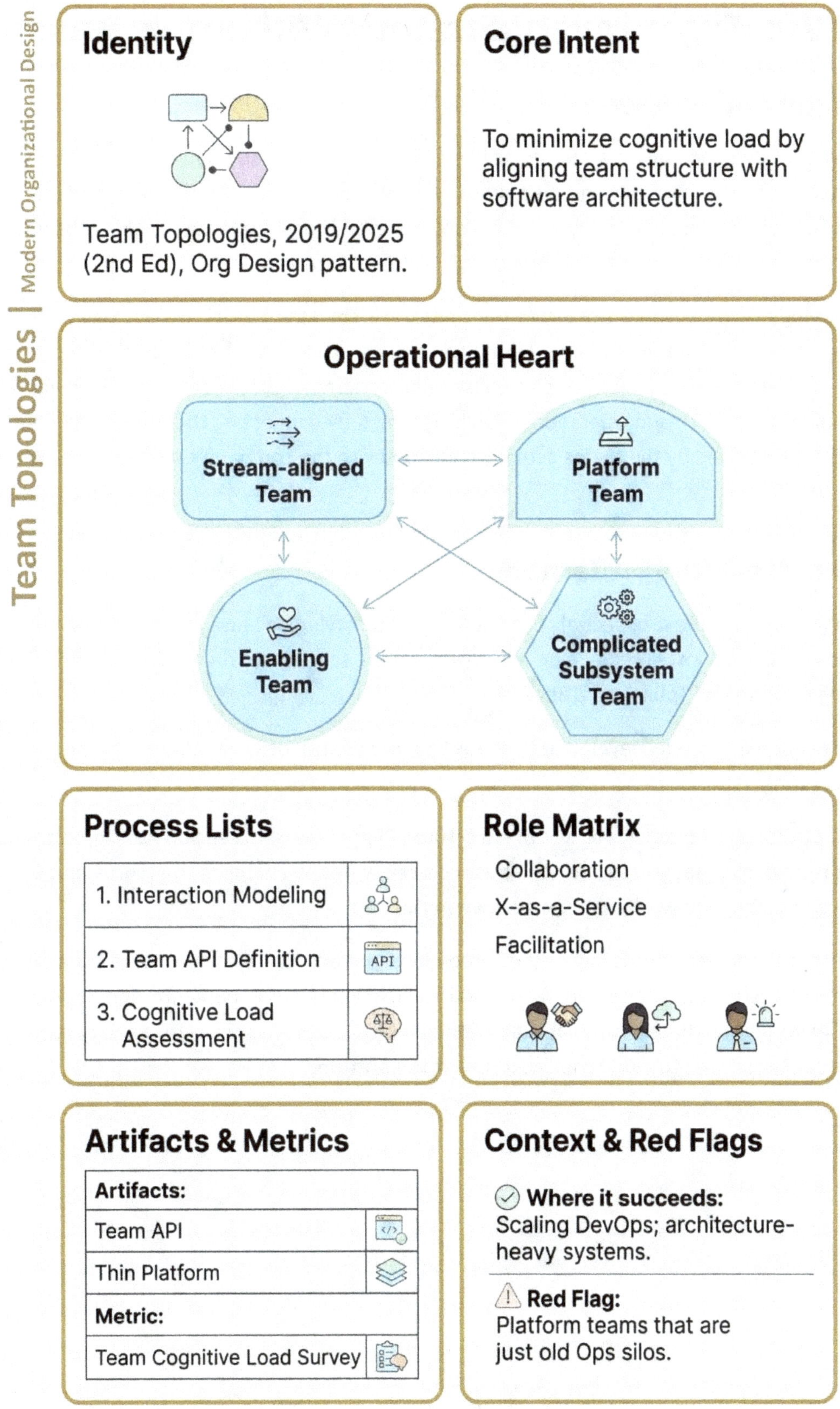

Team Topologies | Modern Organizational Design
Identity
Team Topologies, 2019/2025 (2nd Ed), Org Design pattern.
Core Intent
To minimize cognitive load by aligning team structure with software architecture.
Operational Heart
Stream-aligned Team
Platform Team
Enabling Team
Complicated Subsystem Team
Process Lists
1. Interaction Modeling
2. Team API Definition
API
3. Cognitive Load Assessment
Role Matrix
Collaboration
X-as-a-Service
Facilitation
Artifacts & Metrics
Artifacts:
Team API
Thin Platform
Metric:
Team Cognitive Load Survey
Context & Red Flags
Where it succeeds:
Scaling DevOps; architecture-heavy systems.
Red Flag:
Platform teams that are just old Ops silos.

Figure 19-2: Team Topologies (Modern Organizational Design) Framework Snapshot

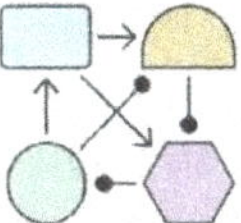

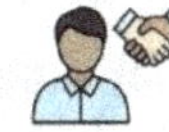

Chapter 20: Flight Levels: Strategy Without Restructuring

20.1. Origin and Ideology

- **The Origin:** Developed by **Dr. Klaus Leopold** (2018), Flight Levels originated from his work in large-scale Kanban transformations and his studies on why Agile teams often fail to impact business agility. Leopold noticed that many companies were optimizing "the team" (making the wheels spin faster) but failing to optimize "the organization" (steering the car).
- **The Ideology:** Flight Levels is not a new method; it is a **Thinking Model**. Its ideology is: "Don't restructure; rewire." It posits that you don't need to fire managers or change job titles to achieve business agility. Instead, you need to visualize how work moves between the Strategy (Level 3), Coordination (Level 2), and Operational (Level 1) layers. By connecting these levels, you ensure that the work being done by the teams actually contributes to the company's strategy.

20.2. The "No-Reorg" Alternative

For many organizations, the prospect of "Scaling Agile" is terrifying because it implies a massive structural upheaval. Frameworks like SAFe or LeSS often require tearing up the org chart, redefining roles, and establishing permanent "Trains" or "Tribes."

Flight Levels, a thinking model developed by Klaus Leopold (2018), offers a different paradigm: **Strategy without restructuring**.

Instead of focusing on *who reports to whom* (structure), Flight Levels focuses on *how information flows* (communication). It posits that you can achieve business agility not by changing the teams, but by visualizing and connecting the work that flows between them.

This approach is particularly appealing to the "predictive" project manager because it allows for improved agility and flow without the trauma and political friction of a full reorganization. It operates on the principle that "Agile Teams" do not automatically create an "Agile Organization". True agility requires connecting the strategy at the top with the execution at the bottom.

20.3. Level 1: Operational (The Team)

Level 1 represents the teams doing the actual work. These are your Scrum, Kanban, or operational teams.

- **The Focus:** Delivery of specific tasks or stories.
- **The Reality:** Most Agile transformations spend 90% of their budget optimizing Level 1. They hire coaches to make the Scrum teams faster.
- **The Trap:** Optimizing Level 1 alone is arguably the most common failure mode in Agile. If a team creates code efficiently (high velocity) but works on the wrong features due to poor strategy, or sits idle waiting for another team, the organization has not improved—it has merely engaged in "local optimization."

20.4. Level 2: Coordination (The Missing Link)

Level 2 is the "end-to-end value stream." It sits between the teams and the strategy.

- **The Focus:** Managing dependencies and flow *between* teams.
- **The Mechanism:** This is often visualized as a "Coordination Board" or a "Program Board." It does not contain tasks; it contains **projects**, **epics**, or **services** that require multiple Level 1 teams to collaborate.
- **Why It Matters:** This is the "missing link" in most organizations. When a project manager asks, *"Why is this feature stuck?"*, the answer is rarely inside a single team—it is usually stuck in the white space *between* teams. Level 2 makes that white space visible.

20.5. Level 3: Strategy (The Portfolio)

Level 3 represents the strategic portfolio. This is where company strategy is connected to execution.

- **The Focus:** Prioritization and outcomes.
- **The Mechanism:** A Strategy Board that limits Work In Progress (WIP) for the entire company.
- **The Shift:** Instead of an annual plan that is "set and forget," Level 3 is an active board where initiatives are killed, paused, or started based on market feedback flowing up from Levels 1 and 2.

20.6. The Interaction: Connecting the Levels

The power of Flight Levels lies not in the boards themselves, but in the specific "Take-off" and "Landing" points that connect them.

- **Strategy Flows Down:** A strategic initiative on the Level 3 board lands on the Level 2 board, where it is broken down into dependencies.
- **Reality Flows Up:** Blockers identified by teams at Level 1 flow up to Level 2 for resolution. If they cannot be resolved, the impact on the timeline flows up to Level 3, triggering a strategic pivot.

By implementing Flight Levels, a project manager moves from being a "taskmaster" to being a "flow manager," ensuring that the organization is not just *busy* (Level 1 activity), but actually *moving* toward its goals.

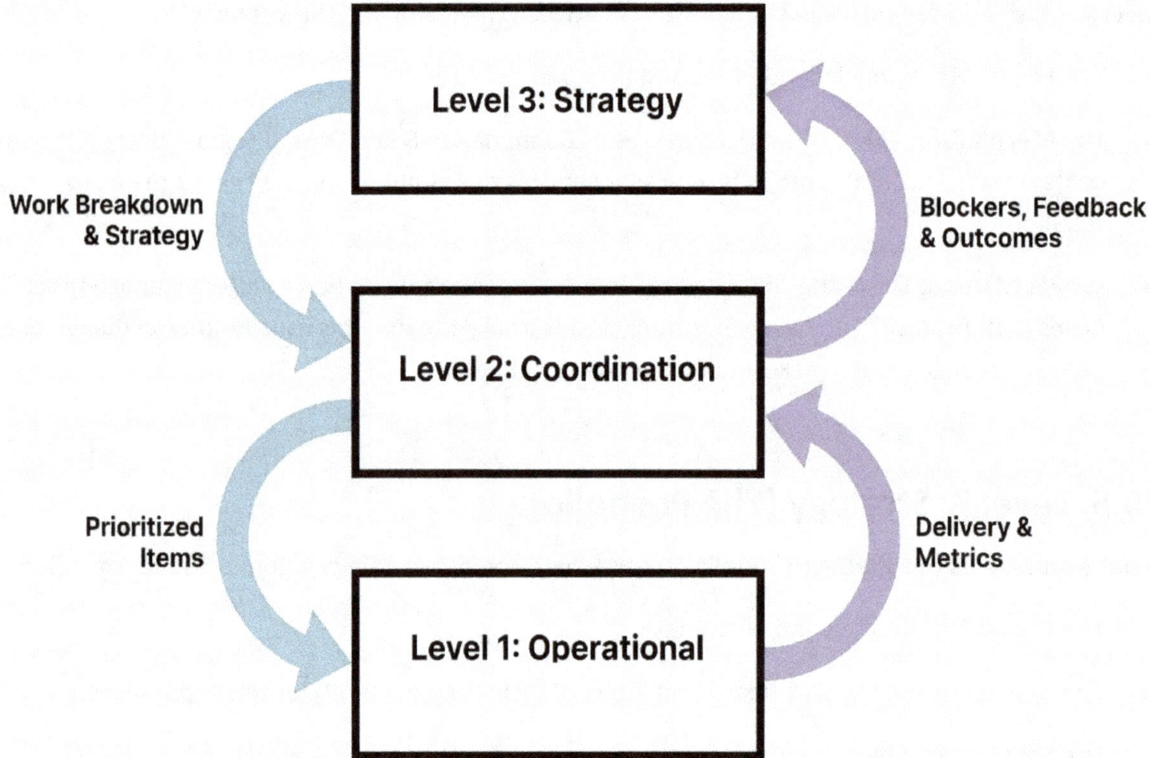

Figure 20-1: Flight Levels: Strategic Information Architecture

20.7. Framework Operational Profile

To provide a concise reference and enable comparison with other frameworks, Flight Levels' practices are summarized in the Operational Profile below.

Table 20-1: Framework Operational Profile: Flight Levels

Feature	Description
1. Agile Approach Category	Strategy Coordination / Business Agility
2. Core Intent / Purpose	To connect organizational strategy to team execution without requiring a full reorganization.
3. Primary Focus	**End-to-End Flow** & **Dependency Management**.
4. Roles	No prescribed roles; uses existing roles (Managers, POs) in new coordination forums.
5. Cadence / Timing Model	Nested Cadences: Strategy reviews (Monthly/Quarterly) connect to Operational syncs (Daily/Weekly).
6. Core Practices	Visualizing the Value Stream, WIP Limits at Strategy Level, Flight Routes.
7. Key Artifacts	Flight Level 2 (Coordination) Boards, Strategy Boards, Topology Maps.
8. Work Management	Flow-based; Strategy flows down, operational reality/data flows up.
9. Primary Metrics	Time-to-Market (End-to-End), Flow Efficiency, WIP at Strategy Level.
10. Organizational Fit	Any organization (startups to enterprises) where team output isn't translating to business results.
11. Strengths	Low resistance to adoption (no re-org required); uncovers hidden bottlenecks between teams.
12. Common Failure Modes	visualizing Level 2 but failing to limit WIP; treating it as just "portfolio management" without connecting to teams.
13. Best Used When...	You have high-performing Agile teams but slow time-to-market due to poor coordination or strategic disconnect.

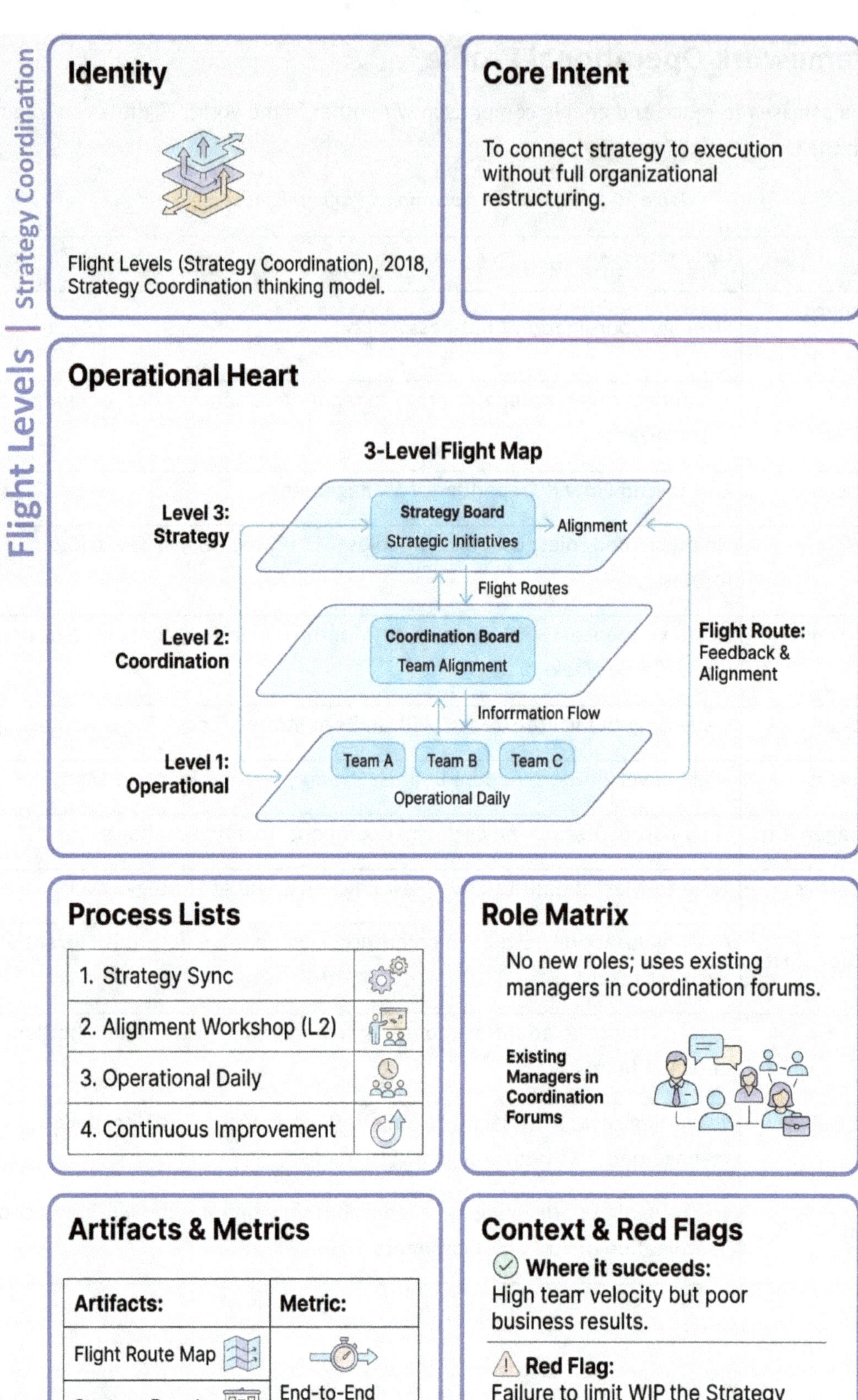

Figure 20-2: Flight Levels (Strategy Coordination) Framework Snapshot

Part V: The Practitioner's Toolkit

Chapter 21: Agile Requirements and Estimation

21.1. User Stories, Epics, and Themes: The INVEST Criteria

Agile requirements are structured hierarchically to maintain both big-picture vision and granular execution.

- **Themes:** High-level strategic objectives (e.g., "Enhance Security").
- **Epics:** Large bodies of work that cannot be completed in a single iteration (e.g., "Implement Multi-Factor Authentication").
- **User Stories:** Small, functional increments written from the perspective of the end user.

Not all teams use all three levels at all times; the hierarchy is a conceptual tool to help maintain alignment between strategy and execution, especially as scale increases.

21.1.1. The INVEST Criteria

To ensure a User Story is ready for development, it should meet the **INVEST** criteria (Wake, 2003)—a cornerstone of high-quality Agile requirements:

- **Independent:** Can be developed and released without a web of dependencies.
- **Negotiable:** It's an invitation to a conversation, not a fixed contract.
- **Valuable:** It must provide clear value to the customer or business.
- **Estimable:** The team understands it well enough to gauge the effort.
- **Small:** Fits within a single iteration (ideally 2–3 days of work).
- **Testable:** Has clear acceptance criteria to determine when it is "Done."

The INVEST criteria should be treated as a guideline rather than a strict checklist. A story may temporarily violate one or more elements during discovery, provided the team understands the risk and addresses it before committing to delivery.

21.2. Backlog Refinement and Just-In-Time Requirements

In traditional projects, requirements are defined upfront (**Big Upfront Requirements**). In *The Agile Landscape*, we use **Just-In-Time (JIT)** requirements through a process called **Backlog Refinement**.

21.2.1. The Refinement "Engine"

Refinement is the ongoing activity of taking "messy" ideas from the bottom of the backlog and sharpening them into "Ready" stories at the top.

- **The Goal:** Ensure the team always has a "Ready" backlog for the next 1–2 iterations.
- **The Benefit:** It prevents the team from over-analyzing features that might be deleted or changed before they are ever built.

Backlog Refinement is not a single meeting but a continuous system that evolves as learning increases and priorities change.

21.3. Estimation: Planning Poker, T-Shirts, and Relative Sizing

In Agile environments, estimation supports decision-making and learning—not contractual certainty. Agile estimation is often misunderstood. We do not estimate to predict the future with 100% accuracy; we estimate to **uncover misunderstandings** and **measure capacity**.

21.3.1. Relative Sizing vs. Absolute Hours

Instead of saying "This will take 14 hours," Agile teams use **Relative Sizing**. Humans are notoriously bad at estimating time but excellent at comparing sizes. We know a "Large" story is roughly twice as big as a "Medium" one, regardless of how many hours they take.

21.3.2. Common Estimation Techniques

1. **Planning Poker:** Using the Fibonacci sequence (1, 2, 3, 5, 8, 13...). This forces the team to acknowledge that as things get bigger, our uncertainty grows.
2. **T-Shirt Sizing (XS, S, M, L, XL):** Best for high-level estimation of Epics or long-term roadmapping.
3. **Affinity Mapping:** A fast way for a team to categorize hundreds of items by placing them in relative "size buckets" on a wall or digital board.

21.3.3. The Real Value of Estimation: From Estimates to Forecasts

The most valuable part of an estimation session is not the resulting number—it's the **discovery of the unknown**. If one developer votes "2" and another votes "13," it signals a massive gap in their understanding of the requirement or the technical complexity. Resolving that gap before work begins is the true purpose of the toolkit. When estimation conversations are skipped or rushed, teams lose an early warning system for hidden complexity and risk.

In the modern landscape, the role of these estimates has evolved. **PMBOK 8 Alignment** highlights a shift from "commitment-based" estimation to **Throughput-Based Flow**. High-maturity teams no longer use Planning Poker to promise a specific date; instead, they use relative sizes to ensure work is small enough to maintain **Value Stream Velocity**.

This historical data is now frequently fed into **Monte Carlo simulations**, which move beyond the fallacy of a single "Target Date" toward a **Probability Range**.

- **Deterministic Estimate (Traditional):** "We will finish this on October 12th."
- **Probabilistic Forecast (Modern):** "Based on our historical velocity and current complexity, there is an 85% chance we will finish this Epic within 4 to 6 weeks."

By shifting from **Deterministic Estimates** to **Probabilistic Forecasts**, the project manager provides the organization with a more resilient view of the delivery ecosystem. This allows for more stable strategic decision-making at the **Portfolio level (Level 3)**, ensuring that the "Strategy Board" is grounded in statistical reality rather than optimistic guesswork.

Chapter 22: Agile Prioritization Techniques

22.1. MoSCoW, the Kano Model, and Buy-a-Feature

Prioritization is rarely about just "gut feeling." Different contexts require different lenses to view value. No single prioritization technique is universally correct; each highlights different aspects of value, risk, and urgency.

22.1.1. MoSCoW: Managing Stakeholder Expectations

As we saw in the DSDM framework (Agile Business Consortium, 2014), MoSCoW is a powerful tool for managing fixed-deadline projects. It categorizes requirements into four buckets:

- **Must have:** Non-negotiable; without these, the product is a failure.
- **Should have:** Important but not vital; can be pained for later if necessary.
- **Could have:** "Nice to haves" that increase satisfaction but aren't critical.
- **Won't have (this time):** Items explicitly deferred to create focus.

MoSCoW works best when "Must haves" are kept intentionally small; overloading this category undermines prioritization entirely.

22.1.2. The Kano Model: Understanding Customer Delight

The **Kano Model** (Kano, 1984) helps teams prioritize features based on how they affect customer satisfaction versus their level of implementation. It identifies three types of features:

1. **Basic Needs (Threshold):** The "taken for granted" features. If they aren't there, the customer is furious; if they are, the customer is neutral (e.g., a "login" button).
2. **Performance Needs:** Satisfaction increases linearly with how well these are implemented (e.g., battery life or search speed).
3. **Delighters (Excitement):** Unexpected features that create high satisfaction and differentiate the product (e.g., an innovative AI feature).

Over time, Delighters often become Performance needs, and Performance needs may become Basic expectations. The Kano categories are therefore dynamic and must be revisited as markets and customer expectations evolve.

22.1.3. Buy-a-Feature: Collaborative Prioritization

This is an "Innovation Game" used to break through stakeholder gridlock. Stakeholders are given a limited amount of "play money" and a list of features, each with a price (based on complexity). Because they cannot afford everything, they must negotiate and pool their money to "buy" the features they value most.

Buy-a-Feature is most effective when facilitated carefully, as power dynamics between stakeholders can otherwise distort the outcome.

22.2. WSJF (Weighted Shortest Job First) and Economic Decision-Making

As organizations scale, prioritization increasingly becomes an economic decision rather than a purely qualitative one. while the previous techniques are excellent for products and stakeholders, **WSJF** (a core component of SAFe; Leffingwell et al., 2023) is the gold standard for large-scale economic prioritization.

22.2.1. The Math of Value

WSJF is designed to solve a specific problem: Should we work on the "Big, High-Value" item first, or the "Small, Medium-Value" item? In a Lean system, the answer is determined by calculating the **Cost of Delay**, a concept applied to product development by Donald Reinertsen (2009).

The formula for WSJF is:

$$WSJF = \frac{Cost\ of\ Delay}{Job\ Size}$$

To calculate the **Cost of Delay**, teams sum three factors:

1. **User-Business Value:** Relative value to the customer.
2. **Time Criticality:** Does the value decay if we don't do it now?
3. **Risk Reduction / Opportunity Enablement:** Does this prevent a future disaster or open a new market?

These inputs are relative estimates, not precise financial calculations; their value lies in comparison, not accuracy.

22.2.2. Why WSJF Works

WSJF prevents the "Loudest Person in the Room" from winning. It favors jobs that deliver the most value in the shortest amount of time. By dividing by **Job Size**, it naturally encourages teams to break work down into smaller, more manageable pieces to increase their priority score.

Practitioner's Tip: Use MoSCoW for day-to-day team focus, the Kano Model for product strategy, and WSJF for high-level roadmap decisions.

Chapter 23: From Velocity to Flow: Evidence-Based Management

23.1. The Trap of Velocity

For years, the primary metric for Agile teams has been **Velocity**: the number of Story Points completed in a Sprint. While useful for internal team capacity planning, Velocity has become a dangerous "vanity metric" in many organizations.

The problem is subjectivity. A "Story Point" is a made-up unit that varies from team to team. Because it is subjective, it is easily manipulated—teams can simply inflate their estimates to appear more productive ("Story Point Inflation"). Furthermore, Velocity measures **output** (how much we built), not **outcome** (how much value we delivered).

In the modern Agile landscape, we move beyond these subjective measures toward objective data derived from **Evidence-Based Management (EBM)** and **Flow Metrics**.

23.2. Evidence-Based Management (EBM)

[Evidence-Based Management (EBM) is a framework that helps organizations measure value rather than just activity. Instead of asking "Did we finish all the tickets?", EBM asks "Did we improve the organization's capability?"

EBM tracks four **Key Value Areas (KVAs)**:

1. Current Value (CV)

- **Question:** Is the product valuable to customers *today*?
- **Metric:** Customer satisfaction (NPS), revenue per employee, or active usage rates. If the team is shipping code but these numbers aren't moving, they are building "waste."

2. Unrealized Value (UV)

- **Question:** What potential market share or value is being missed?
- **Metric:** Market share gap or customer satisfaction gap. This measures the *potential* upside of new features.

3. Time to Market (T2M)

- **Question:** How fast can the organization learn?
- **Metric:** Release frequency and Lead Time. This measures the organization's agility—how quickly it can close the feedback loop.

4. Ability to Innovate (A2I)

- **Question:** Is technical debt preventing new work?
- **Metric:** Defect trends or the ratio of "Innovation Work" vs. "Maintenance Work." If a team spends 80% of its time fixing bugs, its Ability to Innovate is low.

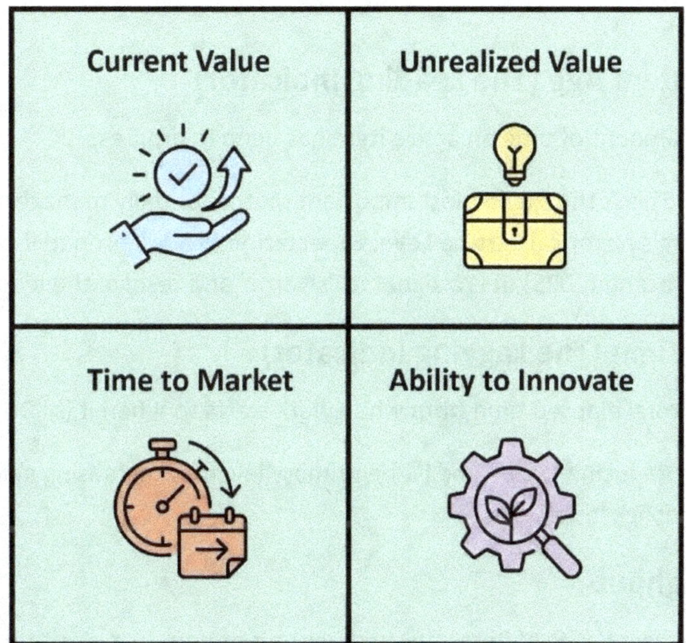

Figure 23-1: Evidence-Based Management (EBM) Quadrants

23.3. Flow Metrics: The Science of Movement

While EBM measures value, **Flow Metrics** measure the efficiency of the delivery process itself. Unlike Story Points, Flow Metrics are objective—they are based on the start and end dates of work items, which cannot be debated.

These metrics are derived from Lean theory and Little's Law and further popularized by Mik Kersten's Flow Framework (Kersten, 2018)

$$WIP = \text{Throughput} \times \text{Cycle Time} \text{ (Also expressed as: } \text{Cycle Time} = \frac{WIP}{\text{Throughput}}\text{)}$$

23.3.1. Work Item Age (The Leading Indicator)

- **Definition:** The amount of time an active item has been in progress.
- **Why it matters:** This is the single most important metric for daily management. If an item's age exceeds the team's historical **Service Level Expectation (SLE)**, a probabilistic forecast detailed in the work of Daniel Vacanti (2025), it is a signal to "swarm" and resolve the blocker immediately.

23.3.2. Cycle Time (The Lagging Indicator)

- **Definition:** The total elapsed time from when work starts to when it is "Done".
- **Why it matters:** Reducing Cycle Time is the primary lever for increasing agility. Faster cycle times mean faster feedback loops.

23.3.3. Throughput

- **Definition:** The exact count of items finished per unit of time (e.g., "5 items per week").
- **Why it matters:** Unlike Velocity, which muddies the water with points, Throughput is a factual count of delivery.

23.4. Probabilistic Forecasting: The End of Estimation

The combination of these metrics allows managers to abandon the 'Crystal Ball' of estimation in favor of **Probabilistic Forecasting** (Vacanti, 2025).

Instead of asking a team to guess how many hours a task will take, modern tools (like Monte Carlo simulations) analyze the team's historical **Throughput** and **Cycle Time** to generate a forecast.

This shifts the conversation from a deterministic lie ("We will be done on Friday") to a probabilistic truth:

*"Based on our historical performance, we have an **85% probability** of completing this specific feature set by November 12th."*

This language appeals directly to the predictive manager's desire for reliability. It acknowledges uncertainty while providing decision-grade data.

23.5. Metric Comparison Summary

Table 23-1: The Shift from Output to Outcome Metrics

Metric Type	**Traditional (Velocity/Points)**	**Modern (Flow/EBM)**
Unit of Measure	Story Points (Subjective)	Days / Items (Objective)
Primary Goal	Predict Capacity	Optimize Flow & Value
Forecasting Method	Team Estimation (Guessing)	Monte Carlo (Historical Data)
Leading Indicator	Burndown Chart	Work Item Age
Management Action	"Why is velocity down?"	"How can we reduce Cycle Time?"

23.6. Measuring Engineering Health: The DORA Metrics

The DORA Metrics Originating from the research of Forsgren, Humble, and Kim (2018), these four metrics have become the gold standard for measuring software delivery performance:

1. **Deployment Frequency:** How often do we release to production? (Throughput)
2. **Lead Time for Changes:** How long does it take for code to go from "commit" to "running"? (Speed)
3. **Change Failure Rate:** What percentage of releases cause a failure? (Quality)
4. **Mean Time to Restore (MTTR):** When failure happens, how quickly do we recover? (Stability) *Note: In 2024/2025, DORA added **Reliability** as a fifth metric, emphasizing that speed without uptime is meaningless.*

Chapter 24: Continuous Improvement

24.1. Retrospectives Beyond Scrum

While the "Sprint Retrospective" is the most famous improvement ritual, the concept of a retrospective belongs to the entire Agile family. In the context of *The Agile Landscape*, a retrospective is not just a meeting—it is a commitment to **psychological safety** and **empirical change**. In mature Agile systems, retrospectives occur at multiple levels—team, product, and organization—to address both local issues and systemic constraints.

24.1.1. The Prime Directive

A successful retrospective begins with Norm Kerth's (2001) Prime Directive, established in his seminal work on project reviews:

"Regardless of what we discover, we understand and truly believe that everyone did the best job they could, given what they knew at the time, their skills and abilities, the resources available, and the situation at hand."

Without this foundation, retrospectives devolve into "Blame-Storming," where teams hide mistakes rather than fixing the system that allowed them to happen.

Psychological safety cannot be mandated by teams alone; it must be actively protected by leadership behavior and organizational incentives.

24.1.2. The Five-Stage Structure

To prevent "Retrospective Fatigue," practitioners use a structured approach to facilitate deep thinking:

1. **Set the Stage:** Prepare the team to engage (e.g., a quick "check-in" on energy levels).
2. **Gather Data:** Create a shared picture of the past cycle using metrics, timelines, or feelings.
3. **Generate Insights:** Ask "Why?" (e.g., using the **5 Whys** technique) to find root causes.
4. **Decide What to Do:** Select **one or two** concrete experiments for the next cycle; with a clear hypothesis and success criteria.
5. **Close:** Summarize and appreciate the team's honesty.

24.2. Kaizen, PDCA (Plan-Do-Check-Act), and Learning Loops

Continuous improvement is not just a team ritual; it is a scientific mindset. Two foundational models from the Lean world provide the backbone for this logic.

Metrics discussed in the previous chapter provide the evidence needed to validate whether improvement experiments are actually working.

24.2.1. Kaizen: Change for Better

Kaizen is the Japanese philosophy of "small, incremental improvements." In an Agile context, Kaizen reminds us that we don't need to revolutionize the organization in a single day. Instead, we aim for "1% better every week."

- **The "Kaizen Event":** Occasionally, a team may pause for a dedicated day or two to solve a deep-rooted architectural or process problem that cannot be fixed during a standard retrospective.

Kaizen is not about working harder; it is about removing friction so that work becomes easier over time.

24.2.2. The PDCA Cycle (The Deming Wheel)

The **Plan-Do-Check-Act (PDCA)** cycle is the scientific method applied to process improvement.

1. **Plan:** Identify an interference or bottleneck and plan a change.
2. **Do:** Carry out the change on a small, controlled scale (the experiment).
3. **Check:** Analyze the data. Did the change help? What did we learn?
4. **Act:** If the experiment was successful, standardize the new way of working. If not, try a different "Plan."

Without the discipline to "Check" and "Act," organizations often remain stuck in perpetual "Plan-Do" cycles that generate activity but little learning.

24.2.3. Single-Loop vs. Double-Loop Learning

In *The Agile Landscape*, we distinguish between two levels of learning:

- **Single-Loop Learning:** Fixing the immediate problem (e.g., "We missed a bug, so let's add a test case").
- **Double-Loop Learning:** Questioning the underlying assumptions (e.g., "Why did we think this feature was necessary in the first place? Is our overall testing strategy flawed?").

True Agility requires teams to move into Double-Loop learning, where they have the courage to challenge not just their tasks, but the very system they work in. This often includes challenging product strategy, funding models, role definitions, or governance structures—not just team practices.

Continuous improvement is the engine that keeps Agile systems alive. Without it, frameworks degrade into rituals, metrics lose meaning, and teams revert to habit-driven behavior. True agility emerges when learning is treated as a first-class outcome, not a side effect.

Part VI: Governance, Leadership, and the Future

Chapter 25: Agile Governance and the Agile PMO

25.1. Agile Contracts, Procurement, and Vendor Management

Traditional procurement is built on the "Fixed Price, Fixed Scope" model. In an Agile world, this creates a paradox: we want to be flexible, but our contracts legally forbid it. To navigate the landscape, we must shift how we buy and sell services. Without changes to contractual incentives, organizations often perform 'Agile theatre' while remaining structurally rigid.

25.1.1. From Fixed-Price to Capacity-Based

Traditional contracts often lead to "adversarial" relationships where vendors hide problems to avoid penalties. Agile contracts focus on **Value and Capacity**:

- **Time and Materials (T&M) with a Ceiling:** Allows for flexibility but protects the buyer from runaway costs.
- **Capped Time and Materials:** The vendor is paid for effort, but there is an agreed-upon "Definition of Done" for specific milestones.
- **Incremental Funding:** Instead of signing a $1M contract, the organization funds "increments" (e.g., $100k per quarter). If the vendor stops delivering value, the funding stops.

These models deliberately shift risk management from contract clauses to ongoing transparency and frequent inspection.

25.1.2. Vendor Management: Partnerships, Not Paychecks

In *The Agile Landscape*, vendors should not be "order takers."

- **Integrated Teams:** Vendor staff should be embedded within internal teams rather than working in a silo.
- **Shared Definition of Done:** Ensure the vendor's quality standards align with your internal standards to avoid "integration nightmares" at the end of the contract.

When vendors are embedded but incentives remain delivery-only, teams risk creating dependency rather than partnership.

25.1.3. The NEC4 Contract (Option C)

For construction and infrastructure, the **NEC4 Contract (Option C)** provides a proven model for Agile collaboration. It uses a "Target Cost" model where the client and contractor share the "pain/gain." If the project comes in under budget, savings are split; if over, overruns are shared. This financially incentivizes collaboration and efficiency, mirroring Agile values within a legally robust framework. (NEC Contracts, 2017)

25.2. From PMO to VMO: The Value Management Office

The traditional Project Management Office (PMO) was often viewed as the "process police," focusing on compliance, standardized templates, and status reporting. Its primary metric was often *conformance to plan*—did we finish on time and on budget? However, in the modern Agile landscape of 2026, this function has undergone a radical evolution into the **Value Management Office (VMO)**.

The VMO shifts the organization's focus from *doing things right* (process adherence) to *doing the right things* (value realization). It does not track "project status" via Red/Amber/Green traffic lights based on arbitrary deadlines; instead, it tracks **Flow Metrics** and **Value Realization**. If a project is "Green" (on schedule) but building a feature customers no longer want, the VMO flags it as a failure of value, not a success of execution. This shift is critical for the predictive manager to understand: the VMO provides *better* control than a traditional PMO because it measures the outcome, not just the output.

25.2.1. From Compliance to Strategic Enablement

The release of the *PMBOK Guide* – Eighth Edition (Project Management Institute [PMI], 2026) formalized this shift, reintroducing structure through Focus Areas while explicitly anchoring them in **Value Delivery Principles**. It validates that governance is not about rigid adherence to a static plan, but about enabling the flow of value.

In this context, the VMO acts as a "Strategy-Execution Bridge." It ensures that the work in the team backlogs actually aligns with the organization's strategic intent (OKRs). The VMO is responsible for **Lean Portfolio Management**, moving the organization away from annual "big bang" funding cycles toward dynamic, quarterly value stream funding. This allows the organization to pivot funding to high-performing initiatives without the bureaucracy of re-approving every individual project charter.

25.2.2. Guardrails, Not Handcuffs

A mature VMO provides governance through **Guardrails** rather than handcuffs. Instead of demanding approval for every decision, the VMO sets clear boundaries within which teams have autonomy.

- **Financial Guardrails:** "You have a budget of $500k for this quarter. You can spend it on any features you deem necessary to achieve the Outcome Goal, provided you do not exceed the cap."
- **Architectural Guardrails:** "You can use any tech stack you want, provided it passes our automated security compliance scans (DevSecOps)."
- **Compliance Guardrails:** Automated checks that ensure regulatory standards (like GDPR or HIPAA) are met without manual inspection.

This "Minimum Viable Bureaucracy" satisfies the organization's need for risk management without stifling the team's speed. It aligns with the **Disciplined Agile philosophy** (PMI, 2022) of being 'compliance-aware' without being "compliance-driven".

25.2.3. Predictive Governance with Agentic AI

Perhaps the most futuristic shift in 2026 is the VMO's adoption of **Agentic AI** to move from *reactive reporting* to *predictive governance*.

In the past, a PMO would report on a delay weeks after it happened. Today, drawing on the capabilities of **Agentic Artificial Intelligence** (Bornet & Wirtz, 2025), the VMO deploys autonomous AI agents that monitor the digital exhaust of the organization—Jira tickets, code commits, and flow metrics—in real time. These agents can detect "Flow Debt" or emerging bottlenecks (e.g., "Team B is blocked by a dependency on Team A, and historical data suggests this will delay the release by 3 weeks") and alert leadership *before* the deadline is missed.

This transforms the VMO from an "Impediment Reporter" into an **"Impediment Destroyer."** Agentic AI doesn't just flag the risk; in advanced VMOs, it can propose solutions, such as reallocating capacity or adjusting scope, allowing the VMO leaders to act as true servant-leaders who clear the path for delivery.

Chapter 26: The Future of Agile

The landscape of work is never static. Just as the Waterfall model evolved into early Agile, the practices of 2025 are rapidly evolving into new forms driven by decentralized teams, artificial intelligence, and modular organizational design.

26.1. Agile in Remote and Hybrid Teams

The "Agile Manifesto" famously prioritizes face-to-face conversation. However, the modern reality is distributed. The future of Agile depends on translating high-bandwidth communication into digital spaces without losing the human connection or drowning in "Zoom fatigue."

- **Asynchronous Agility:** In a world of distributed time zones, the "Daily Stand-up" has evolved from a synchronous meeting into an asynchronous stream of updates. The goal is to create a **Digital Twin** of the team's work that is accessible 24/7.
- **Asynchronous Decision Records (ADRs):** To prevent "decision amnesia" in remote teams, practitioners now use ADRs—lightweight, version-controlled documents that capture the *context*, *rationale*, and *consequences* of a choice. This allows team members to contribute to architectural or strategic pivots without needing to be "in the room" at the same time.
- **The Virtual Obeya:** Physical "War Rooms" are being replaced by infinite digital canvases that serve as the single source of truth for the entire **Value Stream**.
- **Event-Based Co-location:** Hybrid teams are moving toward "Intentional Presence," where teams gather physically not for routine work, but for high-complexity activities like **Product Discovery**, **Strategy Refinement**, or **Deep-Loop Retrospectives**.

Without disciplined transparency and explicit working agreements, remote Agile easily degrades into disconnected execution rather than empowered autonomy.

26.2. The Agentic AI: From Generative to Autonomous

The integration of Artificial Intelligence (AI) is the most significant shift in the history of the movement. We are moving beyond simple automation into the era of **Generative Agile.** AI is no longer just a tool for writing code (like GitHub Copilot); it is becoming an active participant in the management workflow through **Agentic AI**.

26.2.1. Generative vs. Agentic Agile

Between 2023 and 2025, we lived in the era of *Generative AI*—tools that could write code or draft user stories when prompted. In 2026, we have entered the era of **Agentic AI** (Bornet & Wirtz, 2025), where systems do not just "create"; they "act." They can perceive a problem, reason through a workflow, execute multi-step tasks, and learn from the result without constant human intervention. This transition is supported by emerging standards like the **Model Context Protocol (MCP)**, which provides a secure, plug-and-play interface for AI agents to interact directly with enterprise data and project management tools.

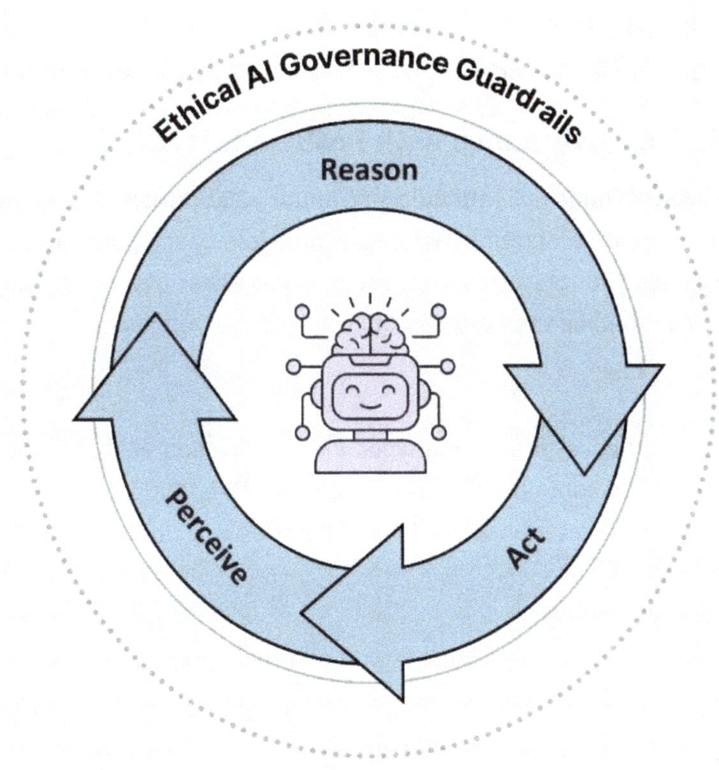

Figure 26-1: The Agentic AI Loop: Perception to Action

26.2.2. The New Team Member: The AI Agent

This fundamentally changes the Agile team structure. We are moving toward **Agentic Teams**, where AI agents are treated as synthetic team members rather than just software tools. This transition requires a new form of team chemistry and accountability.

- **The Human Role:** Humans shift from "Doing" to **"Specifying and Reviewing."** The developer's role becomes that of a **System Architect** who defines the intent, boundaries, and quality standards for the AI agent to execute.
- **The AI Orchestrator:** A critical accountability emerging in 2026 is the **AI Orchestrator**. This person is responsible for the **Systemic Health** of the human-agent ecosystem. Their role is to:
    1. **Choreograph Workflows:** Ensure that multiple agents (e.g., Coding, Testing, and Deployment agents) are synchronized and not conflicting.
    2. **Mitigate Hallucination:** Implement "verification loops" where AI output is cross-checked against the **Definition of Done**.
    3. **Ethical Oversight:** Ensure the agents adhere to the organization's **Sustainability** and **AI Ethics Guardrails**, as mandated by the **PMBOK 8 Governance Performance Domain**.

26.2.3. Predictive Governance with PMBOK 8th Edition

The release of the **PMBOK Guide – Eighth Edition (January 2026)** marks the formal reconciliation of Predictive and Adaptive worlds. Recognizing that purely principle-based guidance (7th Ed) was too abstract for many, the 8th Edition reintroduces structure through **Focus Areas** (Initiating, Planning, Executing, Monitoring, Closing) while retaining the flexibility of Performance Domains.

This is critical for the "Predictive" project manager. PMBOK 8 validates that "process" is not a dirty word in Agile. It provides a structured container (Focus Areas) in which empirical, agentic, and adaptive work can flourish. It effectively ends the "Waterfall vs. Agile" war by offering a unified model for **Value Delivery Systems**.

The Agentic AI | From Generative to Autonomous

Identity

Agentic AI Agile Governance, 2026, Futuristic Governance Model.

Core Intent

To move from reactive status reporting to predictive governance via AI agents.

Operational Heart

Human-AI Hybrid Team

- **Central AI Orchestrator & MCP** (Multi-Agent Control Plane)
- Feedback & Refinement
- **Coding Agent** (Synthetic Teammate)
- **Testing Agent** (Synthetic Teammate)
- **Human-in-the-Loop Review & Approval**

Process Lists

1. Intent Specification
2. Agent Deployment
3. Predictive Alerting
4. Human Review/Approval

Role Matrix

AI Orchestrator

Synthetic Teammates (Agents)

Human Intent Owner

Artifacts & Metrics

Artifacts:	Metric:
MCP Integrations	Value Realization Rate
Real-time Predictive Dashboard	

Context & Red Flags

Where it succeeds:
High digital-exhaust organizations.

Red Flag:
Over-reliance on agents without human intent guardrails.

Figure 26-2The Agentic AI (From Generative to Autonomous) Framework Snapshot

26.3. The unFix Model: The Evolution of Structure

The future of organizational design lies in modularity. The **unFIX Model**, updated in the work of Jurgen Appelo (2025), represents the successor to rigid, "one-size-fits-all" scaling frameworks. Rather than prescribing a static hierarchy, unFIX offers a **Pattern Library** that allows organizations to build their own unique structure.

This approach reinforces the core theme of this book: **Context Counts.** Key patterns include:

- **Crews:** Small, autonomous value-delivery units (Value Stream Crews, Facilitation Crews, or Capability Crews).
- **Bases:** The "home unit" or department where people belong, providing stability and a sense of community even as they move between different Crews.
- **Forums:** Communities of practice that cut across the organization to ensure knowledge sharing and functional excellence.

By treating the organization as a collection of modular patterns, unFIX allows leaders to manage **Cognitive Load** (Chapter 19) through "dynamic re-teaming." It enables the organization to "unfix" rigid silos and "refix" itself around the most current **Value Stream**, providing the ultimate expression of **Adaptive Governance**.

26.4. Agile as an Organizational Capability

The final frontier of *The Agile Landscape* is the realization that "Doing Agile" in IT is insufficient if the rest of the business remains rigid. The future is Business Agility. Without changes to funding models, incentive systems, and decision rights, 'Business Agility' risks becoming a branding exercise rather than a structural capability.

- **Beyond Software:** We are seeing "Agile HR," "Agile Marketing," and "Agile Finance." These departments are adopting the same loops of transparency, inspection, and adaptation to respond to market volatility.
- **From Frameworks to Fluency:** The most successful organizations of the future will not "use" a specific framework like SAFe or LeSS. Instead, they will have developed a high Agile Fluency—the ability to naturally sense a change in the environment and pivot their strategy, funding, and teams within days rather than months.
- **The Talent of the Future:** The most valuable skill in the modern landscape is no longer technical expertise alone, but **Learning Agility**—the ability to unlearn old ways of working and rapidly adapt to new contexts.

Final Reflection

The Agile Landscape is not a map of a static territory; it is a map of a changing world. The frameworks and tools in this book—from the discipline of DSDM to the flow of Kanban and the structural insights of Team Topologies—are your compass. But the journey belongs to you.

Agility is not achieved by adopting practices, but by continuously redesigning the system of work in response to reality.

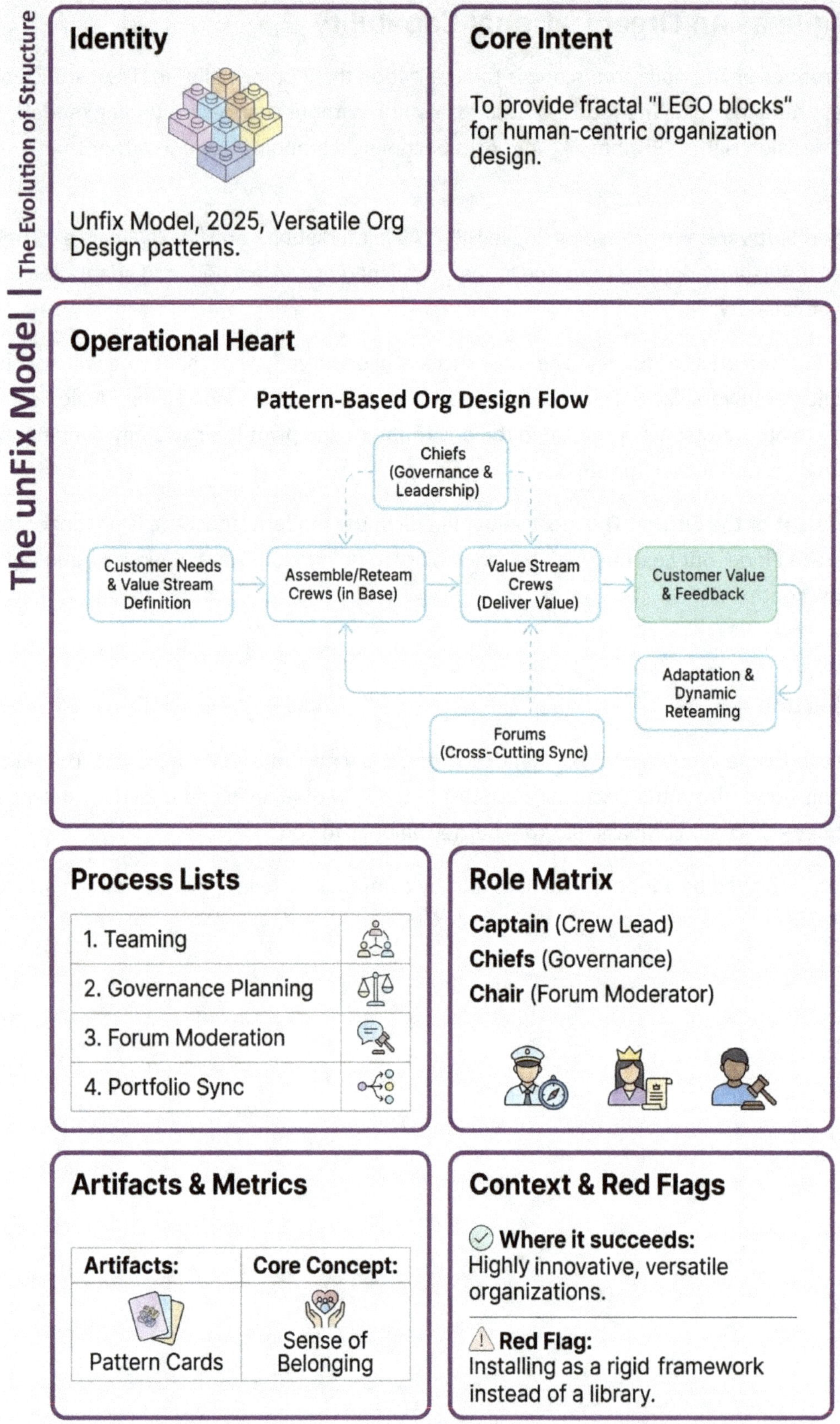

Figure 26-3: The unFIX Model (The Evolution of Structure) Framework Snapshot

References

1. **Adkins, L.** (2010). *Coaching Agile Teams: A Companion for ScrumMasters, Agile Coaches, and Project Managers in Transition*. Addison-Wesley Professional.

2. **Agile Business Consortium.** (2014). *The DSDM Agile Project Framework*. Agile Business Consortium Ltd.

3. **Ambler, S. W.** (2005). The Agile Unified Process (AUP). Scott Ambler + Associates.

4. **Anderson, D. J.** (2010). Kanban: Successful Evolutionary Change for Your Technology Business. Blue Hole Press.

5. **Appelo, J.** (2025). *Human Robot Agent: New Fundamentals for AI-Driven Leadership*. Wiley.

6. **Beck, K.** (2004). Extreme Programming Explained: Embrace Change (2nd ed.). Addison-Wesley Professional.

7. **Bornet, P., & Wirtz, J.** (2025). *Agentic Artificial Intelligence*. World Scientific.

8. **Brooks, F. P., Jr.** (1975). *The Mythical Man-Month: Essays on Software Engineering*. Addison-Wesley

9. **Cascio, J.** (2020). *Facing the age of chaos*. Medium. https://medium.com/@cascio/facing-the-age-of-chaosb00687b1f51d

10. **Cockburn, A.** (2004). Crystal Clear: A Human-Powered Methodology for Small Teams. Addison-Wesley Professional.
11. **Conway, M. E.** (1968). How do committees invent? *Datamation*, 14(5), 28–31.
12. **Dingsøyr, T., Nerur, S., Balijepally, V., & Moe, N. B.** (2012). A decade of agile methodologies: Towards explaining agile software development. Journal of Systems and Software, 85(6), 1213-1221.
13. **Dweck, C. S.** (2006). *Mindset: The new psychology of success*. Random House.

14. **Forsgren, N., Humble, J., & Kim, G.** (2018). *Accelerate: The Science of Lean Software and DevOps*. IT Revolution.

15. **Google.** (2015). *Guide: Understand team effectiveness*. re:Work. https://rework.withgoogle.com/intl/en/guides/understanding-team-effectiveness

16. **ISO/IEC 21031:2024.** *Information technology — Green IT — Software Carbon Intensity (SCI)*. International Organization for Standardization.

17. **Jocham, R., Coleman, J., & Sutherland, J.** (2025). *Scrum Guide Expansion Pack*. ScrumExpansion.org.

18. **Kano, N.** (1984). Attractive quality and must-be quality. *Quality*, 14(2), 39–48.

19. **Kersten, M.** (2018). *Project to Product: How to Survive and Thrive in the Age of Digital Disruption with the Flow Framework*. IT Revolution.

20. **Kerth, N. L.** (2001). Project Retrospectives: A Handbook for Team Reviews. Dorset House Publishing.

21. **Kim, G., & Spear, S.** (2023). *Wiring the Winning Organization: Liberating Our Collective Greatness Through Slowification, Simplification, and Amplification*. IT Revolution.

22. **Ladas, C.** (2009). *Scrumban: Essays on Kanban Systems for Lean Software Development*. Modus Cooperandi Press.

23. **Larman, C., & Vodde, B.** (2016). Large-Scale Scrum: More with LeSS. Addison-Wesley Professional.

24. **Leffingwell, D., et al.** (2023). *SAFe 6.0 Reference Guide: Scaled Agile Framework for Lean Enterprises*. Addison-Wesley.

25. **Lines, S., & Ambler, S. W.** (2020). Choose Your WoW! A Disciplined Agile Delivery Handbook for Optimizing Your Way of Working. Project Management Institute.

26. **Leopold, K.** (2018). *Rethinking Agile: Why Agile Teams Have Nothing To Do With Business Agility*. LEANability Press.

27. **NEC Contracts.** (2017). *NEC4: Engineering and Construction Contract Option C: Target contract with activity schedule*. Thomas Telford Ltd.

28. **Palmer, S. R., & Felsing, J. M.** (2002). *A Practical Guide to Feature-Driven Development*. Prentice Hall.

29. **Project Management Institute.** (2021). A Guide to the Project Management Body of Knowledge *(PMBOK® Guide) – Seventh Edition and The Standard for Project Management*. Project Management Institute.

30. **Project Management Institute.** (2026). *A Guide to the Project Management Body of Knowledge (PMBOK® Guide) – Eighth Edition and The Standard for Project Management*. Project Management Institute.

31. **Project Management Institute.** (2022). *Choose Your WoW! A Disciplined Agile Approach to Optimizing Your Way of Working (Second Edition)*. PMI.

32. **Reinertsen, D. G.** (2009). The Principles of Product Development Flow: Second Generation Lean Product Development. Celeritas Publishing.

33. **Ries, E.** (2011). The Lean Startup: How Today's Entrepreneurs Use Continuous Innovation to Create Radically Successful Businesses. Crown Business.

34. **Schwaber, K., & Sutherland, J.** (2020). *The Scrum Guide: The Definitive Guide to Scrum: The Rules of the Game*. Scrum.org.

35. **Singer, R.** (2019/2025). *Shape Up: Stop Running in Circles and Ship Work that Matters (2nd Edition Update)*. Basecamp.

36. **Skelton, M., & Pais, M.** (2025). *Team Topologies: Organizing Business and Technology for Fast Flow of Value (Second Edition)*. IT Revolution.

37. **Snowden, D. J., & Boone, M. E.** (2007). A leader's framework for decision making. Harvard Business Review, 85(11), 68–76.

38. **Stacey, R. D.** (1996). *Complexity and Creativity in Organizations*. Berrett-Koehler.

39. **Takeuchi, H., & Nonaka, I.** (1986). The new new product development game. *Harvard Business Review*, 64(1), 137–146.

40. **Vacanti, D.** (2025). *Actionable Agile Metrics for Predictability: 10th Anniversary Edition*. ActionableAgile Press.

41. **Wake, B.** (2003). *INVEST in Good Stories, and SMART Tasks*. XP123.

42. **Womack, J. P., & Jones, D. T.** (2003). Lean Thinking: Banish Waste and Create Wealth in Your Corporation. Free Press.

Appendices

Appendix A: Agile Framework Comparison Matrix

Navigating the Landscape

The modern Agile landscape is no longer limited to process frameworks like Scrum or Kanban. It has expanded to include **Product-Led** methodologies (Shape Up), **Organizational Design** patterns (Team Topologies), and **Strategy Coordination** models (Flight Levels).

Note on Selection: This matrix provides a standardized, side-by-side comparison of the **15 primary execution and scaling frameworks** covered in this book. While "The Agile Landscape" explores 18 distinct models in total, **Lean Thinking** is excluded from this specific matrix as it serves as the foundational philosophy underpinning all approaches rather than a standalone delivery framework. Similarly, the **unFIX** Model and **Agentic AI** Governance are excluded here as they represent emerging structural patterns and future-state capabilities rather than established execution methodologies.

Use this guide to move beyond "framework loyalty" and make informed, context-driven decisions for your current organization.

How to Use This Matrix

- **For Selection:** Scan the **"Best Used When..."** and **"Organizational Fit"** columns to shortlist frameworks that align with your team size and problem type (e.g., product discovery vs. service delivery).
- **For Hybridization:** Use the **"Core Practices"** and **"Primary Metrics"** columns to identify specific elements you might blend into your current Way of Working (e.g., adding *Flight Levels* coordination to *Scrum* teams).
- **For Risk Mitigation:** Consult the **"Common Failure Modes"** column to anticipate pitfalls before they occur.

To download a high-resolution PDF of the full matrix, visit:

h-ex-a.com/resources/artifacts/Agile_Framework_Comparison_Matrix

Table 26-1: Agile Framework Comparison Matrix

#	Framework	1. Agile Category	2. Core Intent	3. Primary Focus	4. Roles (Accountabilities)[1]	5. Cadence	6. Core Practices / Ceremonies	7. Key Artifacts
1	**Scrum**	Team-level Framework	Iterative delivery of complex products.	Value Delivery via Empirical Process Control.	PO, Scrum Master, Developers.	Fixed Sprints (1–4 weeks).	Sprint Planning, Daily Scrum, Review, Retrospective.	Product Backlog, Sprint Backlog, Increment.
2	**XP**	Engineering Centric	High-quality code via technical excellence.	Technical Discipline & Reducing Cost of Change.	Customer, Programmer, Coach, Tracker.	Weekly Cycles.	Pair Programming, TDD, CI/CD, Refactoring.	User Stories, Automated Tests, Code.
3	**Kanban**	Lean / Flow	Optimize flow and reduce waste.	Flow Efficiency & Reducing Lead Time.	No prescribed roles (Evolutionary).	Continuous Flow.	Visualize, Limit WIP, Manage Flow, Feedback Loops.	Kanban Board, WIP Limits, CFD.
4	**FDD**	Process Oriented	Scaling through feature-centric modeling.	Domain Modeling & Feature Ownership.	Chief Architect, Chief Programmer, Class Owners.	Feature-based (2-14 days).	Domain Modeling, Develop by Feature, Inspections.	Features List, Object Model, Design Packages.
5	**DSDM**	Governance	Discipline and on-time delivery.	Strategic Alignment & Fixed Constraints.	Sponsor, Visionary, PM, Technical Coordinator.	Fixed Timeboxes.	MoSCoW Prioritization, Timeboxing, Workshops.	Business Case, Prioritized Req. List (PRL), Architecture.
6	**Crystal**	Methodology Family	Human-centric, lightweight delivery.	Communication & Team Safety.	Variable (based on color/team size).	Frequent Delivery.	Osmotic Communication, Reflective Improvement.	Project Map, Release Plan, Status Reports.
7	**AUP**	Hybrid / Legacy	Bridging RUP and Agile.	Lifecycle Coverage (Serial-Large, Iterative-Small).	Modelers, Implementers, Project Manager.	Phased (Inception to Transition).	Model Storming, Test in the Small.	Agile Models, Source Code, Test Suite.
8	**Scrumban**	Hybrid	Transitioning from Scrum to Flow.	Capacity Planning & Flexibility.	Retains Scrum roles (mostly).	Continuous or Sprints.	On-demand Planning, WIP Limits, Daily Sync.	Scrumban Board, Backlog (Trigger-based).

[1] "Roles" is used as a generic term across the landscape, while "Accountabilities" is a specific Scrum nuance.

8. Work Management	9. Primary Metrics	10. Org Fit	11. Strengths	12. Failure Modes	13. Best Used When...
Iterative Pull.	Velocity, Sprint Goal Success.	Small cross-functional teams; complex product environments.	High transparency; clear accountability; rapid risk mitigation.	"Zombie Scrum" (ritual without mindset); "Scrum-but"; Lack of psychological safety.	You need to validate product ideas quickly with a dedicated team in a complex environment.
Iterative Pull (Planning Game).	Code Quality, Test Coverage, Velocity.	Small co-located teams building software with high quality needs.	Lowest defect rates; sustainable pace; high adaptability.	Cultural resistance to pairing; skipping tests under pressure.	Technical risk is high and the cost of software failure is unacceptable.
Continuous Pull.	Cycle Time, Throughput, WIP, SLE.	Service/Support teams, or mature product teams optimizing flow.	Visualizes bottlenecks; handles variability well; low resistance to start.	Over-commitment (ignoring WIP limits); lack of improvement focus.	Work is unpredictable (support/ops) or you want to optimize an existing process without a re-org.
Plan by Feature (Push/Pull hybrid).	Feature Progress (%), Build Health.	Large teams; complex business domains; bank/financial systems.	Scalability; tracking; architectural integrity.	Siloing of class owners; heavy upfront modeling time.	You have a very large team building a complex object-oriented system requiring strict architectural control.
Fixed Time/Cost, Variable Scope.	Value Realization, On-time Delivery.	Corporate/Government with strict fixed deadlines/budgets.	Strong governance; predictability; business engagement.	Too much bureaucracy; treating it as Waterfall.	You are in a corporate/government environment requiring fixed deadlines and auditable governance.
Team-defined.	Delivery Frequency, Morale.	Co-located teams; varies by size (Clear, Yellow, Orange).	Extremely lightweight; high morale; adaptable.	Lack of discipline; failure to reflect and improve.	You need a lightweight, custom approach for a small, co-located team (Crystal Clear).
Phase-based milestones.	Phase Milestones, Defect Trends.	Organizations transitioning from heavy RUP/Waterfall.	Familiar structure for traditional managers; discipline.	Heavy documentation; "Water-Scrum-Fall".	You are transitioning a legacy RUP organization or have heavy regulatory documentation needs.
Pull-based with planning triggers.	Cycle Time, Lead Time.	Maintenance teams; product teams tired of Sprints.	Flexibility of Kanban with structure of Scrum.	Loss of long-term planning; "lazy" Scrum.	Your Scrum team feels constrained by Sprints, or you handle a mix of roadmap and support work.

Table X-1: Agile Framework Comparison Matrix (continue)

#	Framework	1. Agile Category	2. Core Intent	3. Primary Focus	4. Roles (Accountabilities)[1]	5. Cadence	6. Core Practices / Ceremonies	7. Key Artifacts
9	**Shape Up**	Product-Led Value Stream Orchestration	Break the "feature factory" treadmill.	Shaping (Defining) & Betting (Risk Mgmt).	Shapers, Builders (Designers/Devs).	6-Week Cycle + 2-Week Cool-down.	Shaping, Betting Table, Hill Charts, The Circuit Breaker.	The Pitch, Hill Charts, Fat Marker Sketches.
10	**SoS**	Scaling (Basic)	Inter-team coordination.	Synchronization of dependencies.	Ambassador, SoS Master.	Daily or Weekly sync.	The SoS Meeting, Cross-team Retrospective.	Impediment Backlog.
11	**SAFe**	Scaling (Enterprise)	Synchronizing alignment at scale.	Portfolio Alignment & Predictability.	RTE, Product Mgmt, System Arch, SPC.	10-week Planning Interval (PI).	PI Planning, System Demo, Inspect & Adapt.	ART Backlog, Program Board, Arch. Runway.
12	**LeSS**	Scaling (Descaling)	Scrum applied to multiple teams.	Whole Product Focus & Simplification.	1 PO, Scrum Master, Feature Teams.	Synchronized Sprints.	Joint Sprint Planning, Overall Retro, Joint Review.	Single Product Backlog, Integrated Increment.
13	**DA**	Decision Toolkit	Context-driven choice (Choose Your WoW).	Process Tailoring & Flexibility.	Team Lead, Architecture Owner.	Variable (Lifecycle dependent).	Guided Continuous Improvement, Goal Selection.	Way of Working (WoW) agreement.
14	**Team Topologies**	Org Design	Optimize team interactions for flow.	Cognitive Load & Fast Flow.	Stream-aligned, Enabling, Platform, Subsystem.	Evolutionary (Triggered by flow).	Interaction Modes (Collaboration, XaaS, Facilitating).	Team API, Thin Platform.
15	**Flight Levels**	Strategy Model	Connect strategy to execution.	End-to-End Flow across the org.	No prescribed roles (Uses existing).	Nested Cadences (L1, L2, L3).	Visualizing Topology, Strategy Boards, Coordination.	Flight Level Boards (L1, L2, L3).

[1] "Roles" is used as a generic term across the landscape, while "Accountabilities" is a specific Scrum nuance.

8. Work Management	9. Primary Metrics	10. Org Fit	11. Strengths	12. Failure Modes	13. Best Used When...
Betting (No Backlog). Fixed Time, Variable Scope.	Hill Chart Progress, Shipped Bets.	Product companies; startups; mature dev teams.	Focus; no backlog grooming; meaningful work slices.	Shapers disconnecting from reality; "mini-waterfall".	You want to stop backlog grooming and focus on finishing meaningful projects in fixed timeboxes.
Coordinated Team Pull.	Dependency Resolution Rate.	Multiple teams (3-9) on one product.	Lightweight; familiar; low cost.	Becomes a status report; ignores systemic issues.	You have a few teams (3-9) working on a single product and need a lightweight way to connect.
Program-level Pull (PI Objectives).	Flow Predictability, Flow Load.	Global 1000; highly regulated or complex systems.	Alignment; funding/strategy connection; common language.	"SAFe-washing" (bureaucracy); heavy overhead; slow decisions.	You are a Global 1000 enterprise with hundreds of devs needing strict alignment between strategy and execution.
Team Pull from shared backlog.	Value Delivery, Cycle Time.	Product-centric orgs willing to restructure.	Simplicity; customer focus; minimal overhead.	Political resistance to removing management layers.	You want to scale agile by *removing* bureaucracy and organizational complexity (Descaling).
Context-dependent.	G6 Metrics (Value, Quality, etc.).	Organizations with diverse team needs.	Adaptability; agnostic; enterprise awareness.	Analysis paralysis (too many choices); lack of consistency.	You have diverse teams with different needs and want a toolkit to help them optimize their own way of working.
Flow-based; Interaction-driven.	Team Cognitive Load, Flow Efficiency.	Modern software orgs; DevOps/Platform engineering.	Reduces burnout; clear boundaries; enables fast flow.	Creating silos; renaming old teams without changing behavior.	Your teams are burnt out from cognitive overload or blocked by endless hand-offs.
Flow connections between levels.	Time-to-Market, Strategy Execution.	Any org with disconnected strategy and execution.	No re-org needed; connects silos; highlights bottlenecks.	Only visualizing L1 (teams); ignoring L3 (strategy).	You have "Agile Teams" (Level 1) but no "Business Agility" because strategy and coordination are disconnected.

Appendix B: Decision Guide — Selecting the Right Agile Approach

The core premise of this book is that **Context Counts**. Choosing an Agile framework based on popularity rather than fit is the most common cause of transformation failure.

In 2026, the question is no longer just "Scrum or Kanban?" It is a multi-dimensional choice involving **Process**, **Structure**, and **Strategy**. Use this guide to triangulate the right approach for your specific situation.

B.1. The "Choose Your Problem" Selector

Start here. Identify the primary *pain point* your organization is facing.

Table 26-2: Choose Your Problem Selector

If your primary pain point is...	You are solving for...	Consider looking at...	Why?
"We don't know what to build next."	**Discovery & Risk**	**Shape Up (Ch. 13)**	Focuses on "Shaping" the solution and "Betting" resources before the team starts work.
"Our teams are burnt out and blocked by dependencies."	**Cognitive Load**	**Team Topologies (Ch. 19)**	Optimizes team structures to minimize hand-offs and match software architecture (Reverse Conway Maneuver).
"We are busy, but strategy isn't being executed."	**Alignment**	**Flight Levels (Ch. 20)**	Connects strategy (Level 3) to execution (Level 1) without requiring a full organizational restructure.
"We have high technical debt and bugs."	**Quality**	**XP (Extreme Programming) (Ch. 5)**	Provides the engineering discipline (TDD, Pairing) required to sustain agility.
"We need to coordinate 100+ people on one product."	**Scale**	**SAFe (Ch. 16) or LeSS (Ch. 17)**	**SAFe** for structured alignment and portfolio control. **LeSS** for simplification and descaling.

B.2. The Framework Selection Matrix

Use this quadrant to identify a starting point based on **Requirements Uncertainty** and **Delivery Cadence**.

Table 26-3: The Framework Selection Matrix

Context	Recommended Approach	Implementation Logic
High Uncertainty / Fixed Cadence	**Scrum**	You need regular "checkpoints" (Sprints) to validate assumptions and pivot if necessary.
Variable Uncertainty / Continuous Flow	**Kanban**	You cannot predict work 2 weeks ahead (e.g., support/ops). You need to optimize response time (SLA).
High Certainty / Fixed Constraints	**DSDM (AgilePM)**	You have a fixed deadline or budget (e.g., government contract) and need to vary scope (MoSCoW) to hit it.
Product Innovation / Flexible Scope	**Shape Up**	You want to break the "feature factory" cycle and focus on finishing meaningful projects in 6-week bets.

B.3. The Scaling Decision Tree

Before adopting a "heavy" scaling framework, ask these filtering questions:

1. **Do you really need to scale?**
    - *Yes:* We have 50+ people working on *one* single product. -> **Go to Question 2.**
    - *No:* We have 50 people working on *five* different products. -> **Do NOT Scale.** Use separate Scrum/Kanban teams. Use **Flight Levels** for loose coordination.
2. **Is the organization willing to restructure?**
    - *Yes:* We are willing to remove middle management and silos. -> **Choose LeSS (Large-Scale Scrum).** It descales complexity.
    - *No:* We need to map current roles to the new system to maintain stability. -> **Choose SAFe.** It provides a place for existing roles (PMs, Architects) to evolve.
3. **Is the bottleneck "Communication" or "Cognitive Load"?**
    - *Communication:* We just need to talk more. -> **Scrum of Scrums** or **Nexus**.
    - *Cognitive Load:* Teams are overwhelmed by complexity. -> **Team Topologies**. Break the monolith into Stream-aligned and Platform teams.

B.4. Contextual "Red Flags"

Avoid these common mismatches:

- **Avoid Shape Up if:** Your stakeholders cannot refrain from changing requirements for 6 weeks. Shape Up requires a "closed door" during the cycle.
- **Avoid SAFe if:** You are a startup or a small organization (<50 people). The overhead of PI Planning and roles will crush your velocity.
- **Avoid Scrum if:** Your work is driven by interrupts (e.g., a Service Desk). You will constantly fail Sprint goals. Use **Kanban**.
- **Avoid Team Topologies if:** You cannot invest in a "Platform Team." Stream-aligned teams cannot move fast if they have to build their own infrastructure from scratch.

B.5. The "Post-Agile" Assessment

For organizations leveraging AI and Agentic workflows (Chapter 26):

- **If you are deploying Agentic AI:** Shift governance from "Task Verification" to "Outcome Monitoring."
- **Metrics Shift:** Stop tracking *Velocity* (Output). Start tracking *Value Realization* and *Flow Efficiency* (Outcomes).

Appendix C: Agile Glossary of Terms

A

- **Acceptance Criteria:** A set of conditions that must be met before deliverables are accepted. In Agile, these are often defined for each user story to ensure a shared understanding of success.
- **Accountabilities:** A specific term introduced in the 2020 Scrum Guide to replace the concept of "Roles". It shifts focus from job titles to responsible outcomes (e.g., value or effectiveness), reinforcing the team as a single unit without sub-hierarchies.
- **Adaptive Approach:** A development approach in which the requirements are subject to a high level of uncertainty and volatility and are likely to change throughout the project.
- **Affinity Grouping:** A technique used to classify items into similar categories on the basis of their likeness. Often used in Agile for "Affinity Estimating" to size a large backlog quickly.
- **Agentic AI:** An advanced form of artificial intelligence capable of perceiving its environment, reasoning through complex workflows, and executing multi-step tasks autonomously to achieve a high-level goal (e.g., "Resolve this dependency") with minimal human intervention.
- **Agile:** A term used to describe a mindset of values and principles as set forth in the Agile Manifesto. It emphasizes iterative development, customer feedback, and empowered teams.
- **Agile Governance:** An oversight framework that relies on empirical data, transparency, and frequent inspection rather than fixed-plan compliance to ensure alignment and mitigate risk.
- **Agile Release Train (ART):** A long-lived team of Agile teams, which, along with other stakeholders, incrementally develops, delivers, and operates one or more solutions in a value stream.
- **AI Orchestrator:** A new accountability within Agile teams responsible for choreographing the workflows of autonomous AI agents, mitigating hallucinations, and ensuring agents adhere to the "Definition of Done."
- **Anti-pattern:** A common response to a recurring problem that is usually ineffective and risks being highly counterproductive.
- **Appetite:** A constraint used in the *Shape Up* methodology. Instead of estimating how long a task *will* take, the team defines how much time they are *willing* to spend (e.g., "This is a 6-week appetite project").
- **Architecture Owner:** A role (found in **Disciplined Agile**) responsible for guiding the team through architectural and design decisions to ensure long-term technical viability.

- **Augmented Intelligence:** The partnership between human intuition/creativity and AI data processing, where AI suggests options or detects patterns to enhance human decision-making rather than replacing it.

B

- **BANI:** An acronym (Brittle, Anxious, Nonlinear, Incomprehensible) describing the chaotic modern business environment. It replaces or evolves the earlier VUCA model, emphasizing the need for resilience and transparency.
- **Backlog:** An ordered list of work to be done. It serves as the primary source of requirements for any changes to be made to the product.
- **Backlog Refinement:** The progressive elaboration of the content in the backlog and (re)prioritization of it to identify the work that can be accomplished in an upcoming iteration.
- **Base (unFIX Model):** A structural unit in the *unFIX Model* (similar to a "Tribe" or "Business Unit") that provides a home, purpose, and administrative support for multiple Crews.
- **Betting Table:** A meeting in *Shape Up* where leadership allocates resources to pitches for the next cycle. It utilizes a "clean slate" approach, meaning there is no carry-over backlog; unfunded ideas are discarded to prevent hoarding.
- **Big Upfront Requirements (BUFR):** A traditional project management approach where requirements are fully documented before development begins; contrasted with Agile's *Just-In-Time* approach.
- **Bottleneck:** A point in a process where the flow of work is limited or stopped, often visualized in a Cumulative Flow Diagram.
- **Burn-Down Chart** A visual tool used to track the progress of a specific time-bound iteration (such as a Sprint or Cycle) by plotting the remaining work against time. In this book, it is primarily contrasted with the Hill Chart to differentiate between tracking "known" execution tasks and "unknown" discovery phases.
- **Business Agility:** The ability of an organization to sense changes internally or externally and respond accordingly to deliver value to its customers.
- **Business Case:** In an Agile context, a "Lean Business Case" is a lightweight document used to justify an investment based on an initial hypothesis of value, rather than a 50-page speculative plan.

C

- **Cadence:** A regular, predictable rhythm of development events and activities (e.g., Sprints, PI Planning).

- **Capacity-Based Planning:** A method of planning where work is pulled into a period based on the demonstrated historical capacity (velocity or throughput) of the team, rather than being pushed by a deadline.
- **Cognitive Load:** The total amount of mental effort being used in the working memory. In *Team Topologies*, minimizing cognitive load is a primary design principle for software teams to prevent burnout and ensure fast flow.
- **Continuous Guided Improvement (CGI):** The process of using a toolkit to identify and experiment with process improvements based on specific team goals.
- **Continuous Improvement:** The ongoing effort to improve products, services, or processes through small, incremental (Kaizen) or breakthrough improvements.
- **Cost of Delay (CoD):** The economic impact of time on the total value of a project or feature. It is a key factor in Weighted Shortest Job First (WSJF) prioritization.
- **Crew (unFIX Model):** A small, cross-functional team focused on a specific value stream or goal. Equivalent to a "Squad" or "Scrum Team."
- **Cumulative Flow Diagram (CFD):** A chart that shows the status of work items over time, helping to identify Work-in-Progress (WIP) levels and process bottlenecks.
- **Cycle Time:** The amount of time that elapses from the point a work item enters the "In Progress" state until it is completed.
- **Cynefin Framework:** A decision-making framework created by Dave Snowden that categorizes problems into five domains (Clear, Complicated, Complex, Chaotic, and Disorder) to help leaders identify the appropriate response.

D

- **Decentralized Decision-Making:** The practice of pushing decision-making authority down to the lowest possible level of the organization to increase speed and responsiveness.
- **Decision Blade:** A specific process goal in **Disciplined Agile** that provides a list of techniques and their associated trade-offs to help a team "Choose their WoW."
- **Definition of Done (DoD):** A shared understanding of what it means for work to be complete, including quality, testing, and documentation standards.
- **Definition of Outcome Done (DoOD):** A criteria set introduced in the *Scrum Guide Expansion Pack (2025)*. Unlike the standard Definition of Done (which checks if code is built), DoOD checks if the customer behavior has actually changed (value realized).
- **Definition of Ready (DoR):** A team's checklist of criteria that a user story must meet before it can be accepted into a Sprint (e.g., meeting INVEST criteria).

- **Descaling:** The practice of simplifying organizational structure (removing coordination roles, reducing hierarchy) to achieve agility, rather than adding complex processes to manage large numbers of people. A core philosophy of *LeSS*.
- **DORA Metrics:** A set of four (now five) key metrics—Deployment Frequency, Lead Time for Changes, Change Failure Rate, and Mean Time to Recovery (plus Reliability)—used to measure the performance and health of software delivery teams.
- **Double-Loop Learning:** A learning process that involves challenging the underlying assumptions, values, and policies behind a system of work, rather than just fixing a specific error.

E

- **Empiricism:** A process control theory based on transparency, inspection, and adaptation. It is the foundation of Scrum.
- **Epic:** A large, high-level requirement that is too big to be completed in a single iteration and must be broken down into smaller User Stories.
- **Evidence-Based Management (EBM):** A framework that uses empirical data to measure value and organizational capability through four Key Value Areas: Current Value, Unrealized Value, Time-to-Market, and Ability to Innovate.

F

- **Feature Team:** A long-lived, cross-functional team that can complete an end-to-end customer requirement across all layers of the technology stack.
- **Flight Levels:** A thinking model by Klaus Leopold for connecting strategy to execution.
    - **Level 1 (Operational):** The teams doing the work.
    - **Level 2 (Coordination):** Managing dependencies across teams.
    - **Level 3 (Strategy):** Aligning work with enterprise goals.
- **Flow Accelerators:** A set of eight practices in *SAFe 6.0* (e.g., visualizing WIP, working in smaller batches) designed to remove interruptions and speed up value delivery.
- **Flow-Based System:** A delivery model (common in Kanban or Lean) where work is pulled through a system continuously as capacity becomes available, rather than in batches or fixed iterations.
- **Flow Efficiency:** The ratio of active work time (value-add) to total lead time. It highlights how much time work spends "waiting" in queues. High flow efficiency is the primary goal of Lean Portfolio Management.
- **Focus Areas:** Reintroduced in PMBOK Guide 8th Edition, these represent the structural containers of project management (Initiating, Planning, Executing, Monitoring, Closing) applied within a value-delivery context. They replace rigid "Process Groups" to allow for hybrid tailoring.

G

- **Guardrails:** Predefined boundaries or constraints set by governance (often an Agile PMO) within which teams have autonomy to make decisions.

H

- **Hill Chart:** A visual tool used in the Shape Up methodology to track the progress of work based on certainty rather than task completion, divided into the "Uphill" (figuring it out) and "Downhill" (execution) phases.

I

- **Increments** The concrete, usable building blocks of value produced during a delivery cycle. Each increment must meet the "Definition of Outcome Done" (DoOD) and be additive to prior versions, ensuring that value realization is continuous rather than deferred to a single final release.
- **Incremental Funding:** A financial model where budgets are allocated in small, successive tranches based on the delivery of working software and validated value, allowing for "pivoting" or stopping at any point.
- **Information Radiator:** A visible, physical or digital display that provides information to the organization without requiring active questioning.
- **INVEST:** An acronym (Independent, Negotiable, Valuable, Estimable, Small, Testable) used to assess the quality of User Stories.
- **Iteration:** A fixed timebox during which a team produces a valuable, potentially shippable increment of work. Also called a Sprint.

J

- **Just-In-Time (JIT) Requirements:** The practice of refining and detailing requirements only when they are close to being developed, minimizing waste and over-processing.

K

- **Kaizen:** A Japanese philosophy of "change for the better" focusing on continuous, incremental improvement.
- **Kano Model:** A prioritization framework that categorizes features based on customer satisfaction: Basic, Performance, and Delighters.

L

- **Lead Time:** The total time from the moment a request is made (backlog entry) until it is delivered to the customer.

- **Learning Agility:** The ability to unlearn old ways of working and rapidly adapt to new contexts and challenges.
- **Little's Law:** A mathematical principle stating that the average number of items in a system (WIP) is equal to the average arrival rate multiplied by the average time an item spends in the system (Cycle Time). It forms the scientific basis for limiting WIP in Kanban.

M

- **Model Context Protocol (MCP):** A standard allowing AI agents to securely connect with enterprise data sources and tools, enabling them to perform actions (like updating a Jira ticket) rather than just generating text.
- **MoSCoW:** A prioritization technique (Must have, Should have, Could have, Won't have) used to manage scope.
- **Muda, Mura, Muri:** Three Japanese terms from Lean philosophy representing the "three pollutants" of a system: Muda (wasteful activity), Mura (unevenness or inconsistency), and Muri (overburdening people or equipment).

O

- **Objectives and Key Results (OKRs):** A goal-setting framework used to align team efforts with organizational strategy. Often used by an *Agile PMO* to provide strategic guardrails.
- **Obeya:** A Japanese term for "Great Room." A physical or virtual space containing *Information Radiators* used for visual management and coordination.
- **Organizational Agility:** (Often synonymous with Business Agility) The ability of an entire enterprise—not just the IT department—to adapt its structure and strategy to changing market conditions.

P

- **Pair Programming:** An Extreme Programming (XP) practice where two developers work together at one workstation to improve code quality, share knowledge, and reduce errors through continuous review.
- **PI Planning (Program Increment Planning):** A cadence-based, face-to-face (or virtual) event that serves as the heartbeat of the Agile Release Train, aligning all teams on the ART to a shared mission.
- **PDCA (Plan-Do-Check-Act):** Also known as the Deming Cycle. An iterative method used for the control and continuous improvement of processes and products.
- **Planning Poker:** A collaborative estimation technique using the Fibonacci sequence to perform relative sizing of user stories while uncovering misunderstandings.

- **Platform Team:** A team responsible for building and maintaining internal self-service platforms (APIs, tools) that reduce the cognitive load for Stream-aligned teams.
- **Predictive Governance:** The use of AI and flow metrics to forecast project risks and bottlenecks *before* they occur, replacing reactive status reporting.
- **Product Owner (PO):** The role accountable for maximizing the value of the product resulting from the work of the Agile team. They are the sole person responsible for managing the Product Backlog.
- **Process Goal:** A high-level objective in **Disciplined Agile** (e.g., "Explore Scope") that guides teams toward selecting appropriate practices.
- **Psychological Safety:** A shared belief held by members of a team that the team is safe for interpersonal risk-taking. It is the mandatory foundation for any effective retrospective.

Q

- **Queue:** A collection of work items that are waiting to be processed. Managing queue size is a primary focus of Lean-Agile flow.

R

- **Relative Sizing:** An estimation technique where the size of a task is determined by comparing it to other tasks, rather than using absolute time units.
- **Release Train Engineer (RTE):** A servant leader and coach for the Agile Release Train (ART), often described as a "Super Scrum Master."
- **Retrospective:** A regular meeting where the team inspects its process and identifies specific improvements to be implemented in the next cycle.
- **Reverse Conway Maneuver:** The strategic decision to design an organization's communication structure to mirror the desired software architecture. If modular software is desired, the organization must be structured into modular, decoupled teams.
- **Roles:** A generic term for the specific functions and duties assigned within an Agile framework. Unlike traditional hierarchical titles, Agile roles focus on collaborative responsibility and organizational clarity.

S

- **Scrum:** An iterative and incremental framework for developing, delivering, and sustaining complex products.
- **Scrum Master:** The role accountable for establishing Scrum as defined in the Scrum Guide and for the team's effectiveness by enabling the team to improve its practices.

- **Self-Organizing Team:** A team that crosses functional boundaries and manages its own work to achieve a shared goal, rather than being directed by a manager.
- **Servant Leadership:** A leadership philosophy in which the leader's main goal is to serve the team by removing impediments and facilitating growth.
- **Service Level Expectation (SLE):** A probabilistic forecast used in Kanban (e.g., "85% of items are finished in 4 days or less") that replaces deterministic deadlines.
- **Shaping (Shape Up):** The pre-work done by senior leaders to define a solution at the right level of abstraction (solved, but not detailed) before handing it to a team.
- **Stacey Matrix:** A diagram used to assess project complexity by plotting Requirements Uncertainty against Technical Uncertainty to determine if a project is Simple, Complicated, Complex, or Chaotic.
- **Story Point:** A unit-less measure used in relative sizing to represent the overall effort, complexity, and risk of a user story.
- **Stream-aligned Team (Team Topologies):** A team aligned to a single, valuable stream of work (e.g., a specific product feature or user journey), empowered to deliver value independently.

T

- **Team Lead (Agile Lead):** A role in **Disciplined Agile** similar to a Scrum Master, focused on facilitating the team's chosen way of working and removing impediments.
- **Test-Driven Development (TDD):** A technical practice where developers write an automated failing test before writing the actual code, ensuring that the code is testable and meets requirements from the outset.
- **The Agile Landscape:** The conceptual metaphor for the variety of frameworks, tools, and mindsets available in modern practice, emphasizing that "Context Counts."
- **Theme:** A top-level category of work that groups together related Epics to align with high-level strategic objectives.
- **Throughput:** The number of work items (e.g., stories) completed in a specific time period.
- **Timebox:** A fixed period of time during which an activity or event takes place.

U

- **User Story:** A small, brief description of a feature or requirement, told from the perspective of the person who desires the new capability.

V

- **Value Management Office (VMO):** The modern evolution of the PMO. The VMO focuses on optimizing value streams and flow efficiency rather than enforcing compliance with project schedules.
- **Value Stream:** The entire series of activities an organization performs to deliver a product or service to a customer.
- **Velocity:** An indicator of the amount of work a team can handle in a single iteration. It is a capacity planning tool, not a performance metric.
- **VUCA:** An acronym (Volatility, Uncertainty, Complexity, Ambiguity) used to describe the challenging environments projects operate in, highlighting the need for adaptive rather than predictive approaches.

W

- **Way of Working (WoW):** A term popularized by *Disciplined Agile* referring to the specific set of practices a team chooses to follow based on their unique context.
- **Weighted Shortest Job First (WSJF):** A prioritization model used to sequence jobs (e.g., Epics) to maximize value delivery by considering the *Cost of Delay* and *Job Size*.
- **Work-in-Progress (WIP):** The number of work items that have been started but are not yet finished.
- **Work Item Age:** A flow metric measuring the total time an active task has been in progress. It is considered a leading indicator of flow health (unlike Cycle Time, which is a lagging indicator).
- **Working Agreements:** A set of rules or norms created by the team to govern how they will work together and interact.

Index

1–9

A

B

C

D

E

F

G

H

I

J

K

L

M

N

O

P

Q

R

S

T

U

V

W

X

Y

Z

About the Author

M. Rafati is a Project Management Professional (PMP), Architect, and researcher. He holds a Master's degree in Architecture and is the founder of **Hybrid Exploratory Approach**, based in Ontario, Canada.

With a career rooted in the high-stakes fields of Architecture and Construction, he was originally trained in the disciplined landscape of predictive management, where precision and robust risk mitigation are foundational. As the project management field has become increasingly agile, he recognized an urgent need to bridge the gap between traditional governance and modern adaptive flexibility. This realization inspired him to write *The Agile Landscape* as a definitive guide for professionals navigating the transition from rigid, predictive frameworks to the complex, multi-dimensional methodologies of today.

www.ingramcontent.com/pod-product-compliance
Lightning Source LLC
LaVergne TN
LVHW061203120826
845149LV00011B/1890
9781067571306